Quantitative Applications in the Social Sciences

A SAGE PUBLICATIONS SERIES

1. **Analysis of Variance, 2nd Edition** *Iversen/Norpoth*
2. **Operations Research Methods** *Nagel/Neef*
3. **Causal Modeling, 2nd Edition** *Asher*
4. **Tests of Significance** *Henkel*
5. **Cohort Analysis, 2nd Edition** Glenn
6. **Canonical Analysis and Factor Comparison** *Levine*
7. **Analysis of Nominal Data, 2nd Edition** *Reynolds*
8. **Analysis of Ordinal Data** *Hildebrand/Laing/Rosenthal*
9. **Time Series Analysis, 2nd Edition** *Ostrom*
10. **Ecological Inference** *Langbein/Lichtman*
11. **Multidimensional Scaling** *Kruskal/Wish*
12. **Analysis of Covariance** *Wildt/Ahtola*
13. **Introduction to Factor Analysis** *Kim/Mueller*
14. **Factor Analysis** *Kim/Mueller*
15. **Multiple Indicators** *Sullivan/Feldman*
16. **Exploratory Data Analysis** *Hartwig/Dearing*
17. **Reliability and Validity Assessment** *Carmines/Zeller*
18. **Analyzing Panel Data** *Markus*
19. **Discriminant Analysis** *Klecka*
20. **Log-Linear Models** *Knoke/Burke*
21. **Interrupted Time Series Analysis** *McDowall/McCleary/Meidinger/Hay*
22. **Applied Regression** *Lewis-Beck*
23. **Research Designs** *Spector*
24. **Unidimensional Scaling** *McIver/Carmines*
25. **Magnitude Scaling** *Lodge*
26. **Multiattribute Evaluation** *Edwards/Newman*
27. **Dynamic Modeling** *Huckfeldt/Kohfeld/Likens*
28. **Network Analysis** *Knoke/Kuklinski*
29. **Interpreting and Using Regression** *Achen*
30. **Test Item Bias** *Osterlind*
31. **Mobility Tables** *Hout*
32. **Measures of Association** *Liebetrau*
33. **Confirmatory Factor Analysis** *Long*
34. **Covariance Structure Models** *Long*
35. **Introduction to Survey Sampling** *Kalton*
36. **Achievement Testing** *Bejar*
37. **Nonrecursive Causal Models** *Berry*
38. **Matrix Algebra** *Namboodiri*
39. **Introduction to Applied Demography** *Rives/Serow*
40. **Microcomputer Methods for Social Scientists, 2nd Edition** *Schrodt*
41. **Game Theory** *Zagare*
42. **Using Published Data** *Jacob*
43. **Bayesian Statistical Inference** *Iversen*
44. **Cluster Analysis** *Aldenderfer/Blashfield*
45. **Linear Probability, Logit, and Probit Models** *Aldrich/Nelson*
46. **Event History Analysis** *Allison*
47. **Canonical Correlation Analysis** *Thompson*
48. **Models for Innovation Diffusion** *Mahajan/Peterson*
49. **Basic Content Analysis, 2nd Edition** *Weber*
50. **Multiple Regression in Practice** *Berry/Feldman*
51. **Stochastic Parameter Regression Models** *Newbold/Bos*
52. **Using Microcomputers in Research** *Madron/Tate/Brookshire*
53. **Secondary Analysis of Survey Data** *Kiecolt/Nathan*
54. **Multivariate Analysis of Variance** *Bray/Maxwell*
55. **The Logic of Causal Order** *Davis*
56. **Introduction to Linear Goal Programming** *Ignizio*
57. **Understanding Regression Analysis** *Schroeder/Sjoquist/Stephan*
58. **Randomized Response** *Fox/Tracy*
59. **Meta-Analysis** *Wolf*
60. **Linear Programming** *Feiring*
61. **Multiple Comparisons** *Klockars/Sax*
62. **Information Theory** *Krippendorff*
63. **Survey Questions** *Converse/Presser*
64. **Latent Class Analysis** *McCutcheon*
65. **Three-Way Scaling and Clustering** *Arabie/Carroll/DeSarbo*
66. **Q Methodology** *McKeown/Thomas*
67. **Analyzing Decision Making** *Louviere*
68. **Rasch Models for Measurement** *Andrich*
69. **Principal Components Analysis** *Dunteman*
70. **Pooled Time Series Analysis** *Sayrs*
71. **Analyzing Complex Survey Data, 2nd Edition** *Lee/Forthofer*
72. **Interaction Effects in Multiple Regression, 2nd Edition** *Jaccard/Turrisi*
73. **Understanding Significance Testing** *Mohr*
74. **Experimental Design and Analysis** *Brown/Melamed*
75. **Metric Scaling** *Weller/Romney*
76. **Longitudinal Research, 2nd Edition** *Menard*
77. **Expert Systems** *Benfer/Brent/Furbee*
78. **Data Theory and Dimensional Analysis** *Jacoby*
79. **Regression Diagnostics** *Fox*
80. **Computer-Assisted Interviewing** *Saris*
81. **Contextual Analysis** *Iversen*
82. **Summated Rating Scale Construction** *Spector*
83. **Central Tendency and Variability** *Weisberg*
84. **ANOVA: Repeated Measures** *Girden*
85. **Processing Data** *Bourque/Clark*
86. **Logit Modeling** *DeMaris*
87. **Analytic Mapping and Geographic Databases** *Garson/Biggs*
88. **Working With Archival Data** *Elder/Pavalko/Clipp*
89. **Multiple Comparison Procedures** *Toothaker*
90. **Nonparametric Statistics** *Gibbons*
91. **Nonparametric Measures of Association** *Gibbons*

Quantitative Applications in the Social Sciences

A SAGE PUBLICATIONS SERIES

92. **Understanding Regression Assumptions** *Berry*
93. **Regression With Dummy Variables** *Hardy*
94. **Loglinear Models With Latent Variables** *Hagenaars*
95. **Bootstrapping** *Mooney/Duval*
96. **Maximum Likelihood Estimation** *Eliason*
97. **Ordinal Log-Linear Models** *Ishii-Kuntz*
98. **Random Factors in ANOVA** *Jackson/Brashers*
99. **Univariate Tests for Time Series Models** *Cromwell/Labys/Terraza*
100. **Multivariate Tests for Time Series Models** *Cromwell/Hannan/Labys/Terraza*
101. **Interpreting Probability Models: Logit, Probit, and Other Generalized Linear Models** *Liao*
102. **Typologies and Taxonomies** *Bailey*
103. **Data Analysis: An Introduction** *Lewis-Beck*
104. **Multiple Attribute Decision Making** *Yoon/Hwang*
105. **Causal Analysis With Panel Data** *Finkel*
106. **Applied Logistic Regression Analysis, 2nd Edition** *Menard*
107. **Chaos and Catastrophe Theories** *Brown*
108. **Basic Math for Social Scientists: Concepts** *Hagle*
109. **Basic Math for Social Scientists: Problems and Solutions** *Hagle*
110. **Calculus** *Iversen*
111. **Regression Models: Censored, Sample Selected, or Truncated Data** *Breen*
112. **Tree Models of Similarity and Association** *Corter*
113. **Computational Modeling** *Taber/Timpone*
114. **LISREL Approaches to Interaction Effects in Multiple Regression** *Jaccard/Wan*
115. **Analyzing Repeated Surveys** *Firebaugh*
116. **Monte Carlo Simulation** *Mooney*
117. **Statistical Graphics for Univariate and Bivariate Data** *Jacoby*
118. **Interaction Effects in Factorial Analysis of Variance** *Jaccard*
119. **Odds Ratios in the Analysis of Contingency Tables** *Rudas*
120. **Statistical Graphics for Visualizing Multivariate Data** *Jacoby*
121. **Applied Correspondence Analysis** *Clausen*
122. **Game Theory Topics** *Fink/Gates/Humes*
123. **Social Choice: Theory and Research** *Johnson*
124. **Neural Networks** *Abdi/Valentin/Edelman*
125. **Relating Statistics and Experimental Design: An Introduction** *Levin*
126. **Latent Class Scaling Analysis** *Dayton*
127. **Sorting Data: Collection and Analysis** *Coxon*
128. **Analyzing Documentary Accounts** *Hodson*
129. **Effect Size for ANOVA Designs** *Cortina/Nouri*
130. **Nonparametric Simple Regression: Smoothing Scatterplots** *Fox*
131. **Multiple and Generalized Nonparametric Regression** *Fox*
132. **Logistic Regression: A Primer** *Pampel*
133. **Translating Questionnaires and Other Research Instruments: Problems and Solutions** *Behling/Law*
134. **Generalized Linear Models: A Unified Approach** *Gill*
135. **Interaction Effects in Logistic Regression** *Jaccard*
136. **Missing Data** *Allison*
137. **Spline Regression Models** *Marsh/Cormier*
138. **Logit and Probit: Ordered and Multinomial Models** *Borooah*
139. **Correlation: Parametric and Nonparametric Measures** *Chen/Popovich*
140. **Confidence Intervals** *Smithson*
141. **Internet Data Collection** *Best/Krueger*
142. **Probability Theory** *Rudas*
143. **Multilevel Modeling** *Luke*
144. **Polytomous Item Response Theory Models** *Ostini/Nering*
145. **An Introduction to Generalized Linear Models** *Dunteman/Ho*
146. **Logistic Regression Models for Ordinal Response Variables** *O'Connell*
147. **Fuzzy Set Theory: Applications in the Social Sciences** *Smithson/Verkuilen*
148. **Multiple Time Series Models** *Brandt/Williams*
149. **Quantile Regression** *Hao/Naiman*
150. **Differential Equations: A Modeling Approach** *Brown*
151. **Graph Algebra: Mathematical Modeling With a Systems Approach** *Brown*
152. **Modern Methods for Robust Regression** *Andersen*
153. **Agent-Based Models** *Gilbert*
154. **Social Network Analysis, 2nd Edition** *Knoke/Yang*
155. **Spatial Regression Models** *Ward/Gleditsch*
156. **Mediation Analysis** *Iacobucci*
157. **Latent Growth Curve Modeling** *Preacher/Wichman/MacCallum/Briggs*
158. **Introduction to the Comparative Method With Boolean Algebra** *Caramani*
159. **A Mathematical Primer for Social Statistics** *Fox*
160. **Fixed Effects Regression Models** *Allison*
161. **Differential Item Functioning, 2nd Edition** *Osterlind/Everson*
162. **Quantitative Narrative Analysis** *Franzosi*
163. **Multiple Correspondence Analysis** *LeRoux/Rouanet*
164. **Association Models** *Wong*
165. **Fractal Analysis** *Brown/Liebovitch*
166. **Assessing Inequality** *Hao/Naiman*
167. **Graphical Models and the Multigraph Representation for Categorical Data** *Khamis*
168. **Nonrecursive Models** *Paxton/Hipp/Marquart-Pyatt*
169. **Ordinal Item Response Theory** *Van Schuur*
170. **Multivariate General Linear Models** *Haase*

Series/Number 09-170

MULTIVARIATE GENERAL LINEAR MODELS

Richard F. Haase
University at Albany, State University of New York

SAGE

Los Angeles | London | New Delhi
Singapore | Washington DC

SAGE

Los Angeles | London | New Delhi
Singapore | Washington DC

FOR INFORMATION:

SAGE Publications, Inc.
2455 Teller Road
Thousand Oaks, California 91320
E-mail: order@sagepub.com

SAGE Publications Ltd.
1 Oliver's Yard
55 City Road
London EC1Y 1SP
United Kingdom

SAGE Publications India Pvt. Ltd.
B 1/I 1 Mohan Cooperative Industrial Area
Mathura Road, New Delhi 110 044
India

SAGE Publications Asia-Pacific Pte. Ltd.
33 Pekin Street #02-01
Far East Square
Singapore 048763

Acquisitions Editor: Vicki Knight
Associate Editor: Lauren Habib
Editorial Assistant: Kalie Koscielak
Production Editor: Brittany Bauhaus
Copy Editor: QuADS PrePress (P) Ltd.
Typesetter: C&M Digitals (P) Ltd.
Proofreader: Lori Newhouse
Indexer: Diggs Publishing Services, Inc.
Cover Designer: Candice Harman
Marketing Manager: Helen Salmon
Permissions Editor: Adele Hutchinson

Library of Congress Cataloging-in-Publication Data

Haase, Richard F.
Multivariate general linear models/
Richard F. Haase.

p. cm.
(Quantitative applications in the social sciences; 170)
Includes bibliographical references and index.

ISBN 978-1-4129-7249-9 (pbk.: alk. paper)

1. Social sciences—Statistical methods.
2. Multivariate analysis. 3. Regression analysis.
I. Title.

HA31.35.H33 2011
519.5′35—dc23 2011032017

This book is printed on acid-free paper.

11 12 13 14 15 10 9 8 7 6 5 4 3 2 1

CONTENTS

ABOUT THE AUTHOR

Richard F. Haase is Professor Emeritus and Research Professor in the Division of Counseling Psychology of the School of Education and Fellow of the Institute for Health and the Environment of the School of Public Health, both at the University at Albany, State University of New York. After completing his PhD in psychology from Colorado State University, he has taught research methods, statistics, and data analysis at the University of Massachusetts at Amherst, Texas Tech University, and the University at Albany. His interests are in the areas of research methods, univariate and multivariate statistics, and vocational psychology. His work on research methodology and data analysis has appeared in the *Journal of Consulting and Clinical Psychology*, *Journal of Counseling Psychology*, *Educational and Psychological Measurement*, *Multivariate Behavioral Research*, *Applied Psychological Measurement*, *Environmental Research*, and the *Journal of Vocational Behavior*.

CHAPTER 1. INTRODUCTION AND REVIEW OF UNIVARIATE GENERAL LINEAR MODELS

Few data analytic techniques command a position of greater importance in the social, behavioral, and physical sciences than multiple regression analysis. Exemplary applications can be found in the full range of disciplines, including anthropology (Cardoso & Garcia, 2009), economics (Card, Dobkin, & Maestas, 2009), political science (Baek, 2009), sociology (Arthur, Van Buren, & Del Campo, 2009), and all branches of psychology (Ellis, MacDonald, Lincoln, & Cabral, 2008; Pekrun, Elliot, & Maier, 2009).

In each of these disciplines, the purpose of the investigator is to study the relationship between the variables. Fitting regression models to data allows the analyst the ability to account for or explain variation in a criterion variable as a function of one or more predictor variables. The general linear model is an extension of regression models to accommodate both qualitative and quantitative predictor variables. It is widely recognized that multiple regression analysis is a data analytic system that subsumes all linear models (Cohen, 1968), including those that are based on continuously distributed predictor variables (classic regression analysis), those that are based on schemes to accommodate categorical predictors (classic analysis of variance), and those models that are based on any combination of continuous and categorical predictors.[1] Together these models define the general linear model. The regression model is flexible enough to handle many different realizations of predictor variables, including interactions between continuous predictor variables, between categorical predictor variables, and between combinations of continuous and categorical predictor variables. The breadth of coverage of possible analyses afforded by these combinations explains why the technique is so widely used in all scientific disciplines from anthropology to zoology.

In this volume, our goal is to introduce the multivariate version of the general linear model and to illustrate several of its applications. Multivariate models are distinguished by the presence of more than one dependent

[1]Some authors prefer the terms *quantitative* and *qualitative* to describe predictor variables that are continuous or categorical. In this volume, we use the term *continuous* to denote variables whose underlying metric is continuous or discrete, and we use the term *categorical* to denote nominal group structure that has no meaningful underlying metric except to identify categories.

variable that are to be analyzed simultaneously by fitting a single model to the data. Much of the conceptual and statistical basis of multivariate linear model analysis is a direct generalization of univariate regression analysis, which we briefly review in this chapter. This review of univariate strategies for analyzing linear models is intended to set the stage for the remaining chapters. In Chapter 2, we introduce the example data sets to be used throughout along with a discussion of the first step in the general linear model (GLM) analysis of specifying the model. In Chapters 3, 4, and 5, we cover the estimation of parameters of the model, the assessment of goodness of fit of the model along with the related multivariate test statistics, and testing hypotheses on the model. Chapter 6 introduces the linear model solution to the multivariate analysis of variance, and Chapter 7 concludes the volume with an introduction to canonical correlation analysis, which is a linear model that subsumes all of the material of preceding chapters. The overriding goal of the text is to present an integrated view of all these various techniques under a single modeling framework.

Review of Univariate Linear Model Analysis

The main goal of the linear model is to evaluate relationships in order to explain variability in a response variable as a function of some specified model and an error of prediction:

$$\text{Response} = \text{Model} + \text{Error}.$$

In the *univariate* case, regression models are those models that are limited to a single criterion, response, dependent, or outcome variable.[2] Univariate regression models can be expressed mathematically as a regression function,

$$Y = \beta_0 + \beta_1 X_1 + \varepsilon, \qquad [1.1]$$

[2]We use the terms *dependent*, *criterion*, *response*, and *outcome* interchangeably in this volume to describe the *Y* variable in models. The *X* variables in the model will be interchangeably referred to as predictor, explanatory, or independent variables. These terms appear throughout the literature on regression analysis. Some authors prefer to reserve the term *dependent variable* to experimental designs with manipulated conditions.

for a simple model with a single predictor variable. For a more complex model with multiple predictors, we may write[3]

$$Y = \beta_0 + \beta_1 X_1 + \beta_2 X_2 + \cdots + \beta_q X_q + \varepsilon. \quad [1.2]$$

In Equations 1.1 and 1.2, Y represents a single column vector response variable that is intended to be explained by the weighted linear combination of regression coefficients, $\beta_0, \beta_1, \ldots, \beta_q$, and explanatory variables, $X_1, X_2, \ldots, X_q$, and includes a disturbance or error term ε, which captures all other sources of variability, both systematic and random, that are responsible for variation in Y. The X_j explanatory variables, $j = 1, 2, \ldots, q$, can be either continuous or categorical.[4] Many contemporary textbooks emphasize this integrative linear model approach to both regression analysis and the analysis of variance in the univariate case (see, e.g., Cohen, Cohen, West, & Aiken, 2003; Myers & Well, 2003).

Although we briefly review the basic ideas of univariate regression/linear model analysis in this chapter, our purpose is to set the stage for the analysis of *multivariate* multiple regression/general linear model analysis with continuous and categorical predictor variables—multivariate models can be conceptualized as generalizations of their univariate counterparts. Whereas univariate regression models are defined by their single column vector of Y scores, multivariate models are defined largely by the fact that *more than one dependent variable* is simultaneously included in the model specification. The collection of the explanatory variables, $X_1, X_2, \ldots, X_q$, can be identical for univariate and multivariate models; only the number of Y variables, the number of columns of regression coefficients, and the number of associated disturbance terms, ε, will differ.

As models become more complex, it will be convenient to express the models and their applications in matrix algebraic terms. Although we introduce the basic matrix notation to identify the linear models discussed in this volume, we do not present a full coverage of the topic. A chapter-length coverage of many of the details is given in Draper and Smith (1998,

[3]We do not identify the response and explanatory variables Y or X with a subscript to indicate the serial order of the 1st through the nth observations. In this volume, all models are based on the full set of n observations, and the index of summation or multiplication is assumed to be across all n participants.

[4]Coding schemes for categorical variables will be introduced at greater length in later sections.

Chap. 4); textbook-length coverage can be found in Namboodiri (1984) or Schott (1997).

The univariate multiple regression model of Equations 1.1 and 1.2 can be conveniently summarized in matrix notation as[5]

$$\mathbf{y}_{(n\times 1)} = \mathbf{X}_{(n\times q+1)}\,\boldsymbol{\beta}_{(q+1\times 1)} + \boldsymbol{\varepsilon}_{(n\times 1)} \qquad [1.3]$$

in which $\mathbf{y}_{(n\times 1)}$ is a single-column vector whose *dimensions* are noted in the row-by-column subscript. The X_j predictor variables, $j = 1, 2,\ldots, q$, collected in a design matrix, $\mathbf{X}_{(n\times q+1)}$, are the counterpart of the same predictor variables in the univariate model of Equation 1.2, now expressed as a matrix of *order* $(n \times q + 1)$ with n rows identifying each of the $i = 1, 2, \ldots, n$ cases and $q + 1$ columns that capture the predictor variables. The "+1" in the $q + 1$ dimension allows for a unit vector of $X_0 \equiv 1$ ($\equiv$ means "by definition equal to") to estimate the intercept of the model. The vector $\boldsymbol{\beta}$ of Equation 1.3 is a $(q + 1 \times 1)$ column vector of regression coefficients containing one row for each of the $q + 1$ explanatory variables. Expanding Equation 1.3 shows the elements contained in the matrices for a univariate multiple regression model with $q + 1$ predictor variables:

$$\begin{bmatrix} Y_1 \\ Y_2 \\ \vdots \\ Y_n \end{bmatrix} = \begin{bmatrix} 1 & X_{11} & X_{12} & \cdots & X_{1q} \\ 1 & X_{21} & X_{22} & \cdots & X_{2q} \\ \vdots & \vdots & \vdots & \ddots & \vdots \\ 1 & X_{n1} & X_{n2} & \cdots & X_{nq} \end{bmatrix} \begin{bmatrix} \beta_0 \\ \beta_1 \\ \vdots \\ \beta_q \end{bmatrix} + \begin{bmatrix} \varepsilon_1 \\ \varepsilon_2 \\ \vdots \\ \varepsilon_n \end{bmatrix}.$$

The multivariate multiple regression model is a generalization of Equation 1.3 and would be written as

$$\mathbf{Y}_{(n\times p)} = \mathbf{X}_{(n\times q+1)}\,\mathbf{B}_{(q+1\times p)} + \mathbf{E}_{(n\times p)}. \qquad [1.4]$$

The matrix $\mathbf{Y}_{(n\times p)}$ is a two-dimensional array of numbers in which the rows of the matrix represent all the n observations (subjects, cases) and the columns of the matrix contain the $p > 1$ response variables, Y_k, for $k = 1, 2, \ldots, p$. Hence, the *order* of the matrix $\mathbf{Y}$ is $(n \times p)$. The structure

[5]We use italics to represent scalars (e.g., $X, Y, Z, \beta, \varepsilon$), boldface lowercase letters to denote row or column vectors (e.g., $\mathbf{a}, \mathbf{b}, \mathbf{y}, \mathbf{x}, \boldsymbol{\beta}, \boldsymbol{\varepsilon}$), and boldface uppercase letters to denote matrices (e.g., $\mathbf{X}, \mathbf{Y}, \mathbf{B}, \mathbf{E}, \boldsymbol{\Gamma}$). If a column or row vector is deliberately represented by a matrix symbol, its vector status will be made explicit by the order of the matrix, e.g., $(n \times 1)$ or $(1 \times p)$.

of the design matrix, $\mathbf{X}_{(n \times q+1)}$, does not differ from univariate to multivariate models and is identical to that of Equation 1.3. The matrix $\mathbf{B}_{(q+1 \times p)}$ of Equation 1.4 is an augmented collection of regression coefficients, one row for each of the $q + 1$ explanatory variables and p columns to accommodate the multiple response variables. Finally, the matrix $\mathbf{E}_{(n \times p)}$ is a collection of vectors of disturbance terms, one row for each of the n cases on each of the p response variables in the model. Expanding Equation 1.4 reveals the matrix elements that would be contained in the multivariate model,

$$\begin{bmatrix} Y_{11} & Y_{12} & \cdots & X_{1p} \\ Y_{21} & Y_{22} & \cdots & Y_{2p} \\ \vdots & \vdots & \ddots & \vdots \\ Y_{n1} & Y_{n2} & \cdots & Y_{np} \end{bmatrix} = \begin{bmatrix} 1 & X_{11} & X_{12} & \cdots & X_{1q} \\ 1 & X_{21} & X_{22} & \cdots & X_{2q} \\ \vdots & \vdots & \vdots & \ddots & \vdots \\ 1 & X_{n1} & X_{n2} & \cdots & X_{nq} \end{bmatrix}$$

$$\begin{bmatrix} \beta_{01} & \beta_{02} & \cdots & \beta_{0p} \\ \beta_{11} & \beta_{12} & \cdots & \beta_{1p} \\ \vdots & \vdots & \ddots & \vdots \\ \beta_{q1} & \beta_{n2} & \cdots & \beta_{qp} \end{bmatrix} + \begin{bmatrix} \varepsilon_{11} & \varepsilon_{12} & \varepsilon_{1} & \varepsilon_{n1} \\ \varepsilon_{21} & \varepsilon_{22} & \varepsilon_{2} & \varepsilon_{n2} \\ \vdots & \vdots & \ddots & \vdots \\ \varepsilon_{n1} & \varepsilon_{n1} & \varepsilon_{n} & \varepsilon_{np} \end{bmatrix}$$

In the succeeding chapters, we will pursue more of the details of structuring the design matrix to accommodate both continuous and categorical predictor variables. For the remainder of this chapter, we set the stage by focusing on a review of univariate linear models.

We assume that the reader has a reasonably good understanding of univariate multiple regression analysis at the level of Cohen et al. (2003) and a similarly good understanding of analysis of variance models at the level of Myers and Well (2003). We also assume an elementary grasp of matrix addition, subtraction, multiplication, and inverse (division). We hope to show that much of multivariate analysis can be seen as a generalization of univariate analysis. Toward that end, we turn now to a review of the univariate regression model in which we introduce four steps of general linear model analysis:

1. Specify the model.

2. Estimate the parameters of the model.

3. Define measures of goodness of fit of the model.

4. Develop methods for testing hypotheses about the model.

Because of space constraints, we do not undertake a discussion of diagnosis of the adequacy of the models that is covered in detail elsewhere (Cohen et al., 2003, Chap. 4).

Specifying the Univariate Regression Model

The dimensions of $\mathbf{Y}_{(n \times p)}$ define the initial distinction between univariate and multivariate models. If the designation of the model includes a single-column vector of scores, then $\mathbf{y}_{(n\times 1)}$ represents the dependent variable as noted in Equation 1.3. Consider a regression model in which $\mathbf{y}_{(n\times 1)}$ is hypothesized to be a function of three predictors—continuously distributed variables X_1 and X_2 and a dichotomous categorical variable X_3. Ultimately data must be collected that conform to the model specifications. To make matters more concrete, let Y represent the construct of executive functioning as measured by scores on the Trail Making Test–Part B (TMT-B, Tombaugh, 2004). Neuropsychologists consider the TMT-B to be a measure of higher-order brain function governing the activities of planning, organization, and anticipation. Since executive functioning is a critical cognitive skill, understanding how status on this dimension might vary with advancing age, increasing education, and differences in gender is important. A fictitious data set based on $n = 40$ observations with correlation structure nearly identical to that reported by Tombaugh (2004) specifies a three-predictor model defined in Equation 1.3. The prototypical matrices required to specify this linear model would include the following:

$$\mathbf{y}_{(40\times 1)} = \begin{bmatrix} 72 \\ 115 \\ 117 \\ \vdots \\ 111 \end{bmatrix}, \mathbf{X}_{(40\times 4)} = \begin{bmatrix} 1 & 41 & 13 & 0 \\ 1 & 51 & 18 & 1 \\ 1 & 80 & 14 & 0 \\ \vdots & \vdots & \vdots & \vdots \\ 1 & 59 & 10 & 0 \end{bmatrix}, \boldsymbol{\beta}_{(4\times 1)} = \begin{bmatrix} \beta_0 \\ \beta_1 \\ \beta_2 \\ \beta_3 \end{bmatrix}, \boldsymbol{\varepsilon} = \begin{bmatrix} \varepsilon_1 \\ \varepsilon_2 \\ \varepsilon_3 \\ \vdots \\ \varepsilon_{40} \end{bmatrix}.$$

In this univariate model, the vector $\mathbf{y}$ is time to completion of the TMT-B task, X_1 is the participant's age and X_2 is the participant's education, both continuous predictor variables. The vector X_3 is a dummy-coded regressor, representing a categorical variable of gender coded as 1 = female and 0 = male. The vector $X_0 \equiv 1$ is included as the first column of the design matrix to accommodate the model intercept. The means, standard deviations, and correlations for these data are shown in Table 1.1.

Articulating this descriptive information along with writing out the regression model specified in Equation 1.2 or 1.3 are the statistical details required to specify the model.

Table 1.1 Means, Standard Deviations, and Correlations for the TMT-B Data

	TMT-B	*Age*	*Education*	*Gender*
TMT-B	1.000			
Age	.632	1.000		
Education	–.244	–.171	1.000	
Gender	–.046	.014	–.114	1.000
Mean	93.77	58.48	12.60	.45
Standard deviation	32.77	21.68	2.60	.50

Note: $n = 40$. TMT-B = Trail Making Test–Part B.

A second important aspect of linear model specification depends heavily on the theory that dictates the mathematical model and provides the substantive explanation of the hypothesized relationship between response and explanatory variables. The theoretical basis of the research often includes the logic used to explain the mechanism through which the Y and X variables are presumed to be associated. These very important details of model specification are context specific and will vary from study to study. While we will endeavor to provide the flavor of such arguments in the examples used to illustrate the procedures here, a full discussion of this aspect of model specification is beyond the scope of this volume. Extensive coverage of this topic is given in Jaccard and Jacoby (2010).

Estimating the Parameters of the Model

The models of Equations 1.1 to 1.4 are population regression functions with parameters of the model defined in the elements of $\boldsymbol{\beta}_{(q+1\times 1)} = (\beta_0, \beta_1, \ldots, \beta_q)$ for univariate models and of $\mathbf{B}_{(q+1\times p)}$ for the multivariate case. For the q-predictor univariate regression model of Equation 1.3, it is known that the long-run expected value of the function for a single criterion variable is given by

$$E(Y|\mathrm{X}) = \mathbf{X}\boldsymbol{\beta} = \beta_0 + \beta_1 X_1 + \beta_2 X_2 + \cdots + \beta_q X_q. \qquad [1.5]$$

Figure 1.1 The Linear Regression Function With Expected Values (Means) of the Conditional Distributions of Y on X for the Data of Table 1.1

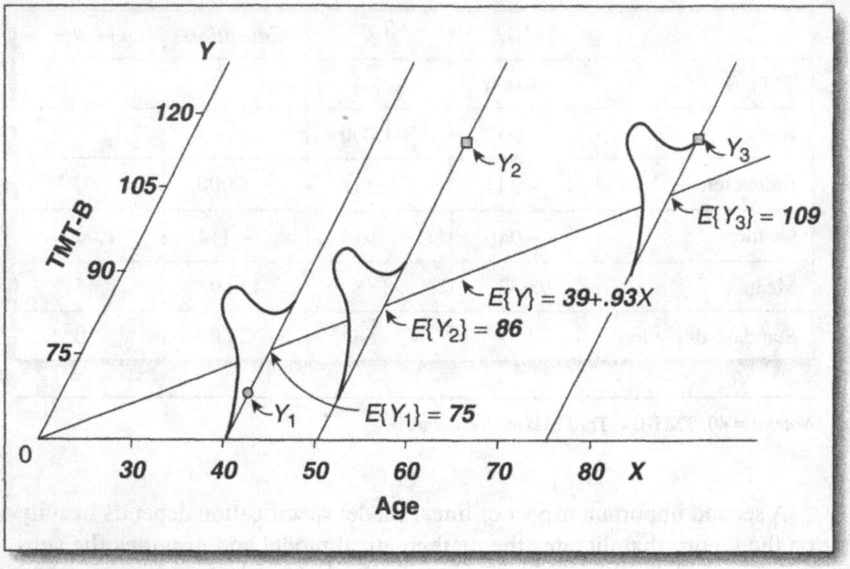

Note: Y_1, Y_2, and Y_3 are illustrative cases.

These expected values are the means of the conditional probability distributions of Y, say $\mu_{(Y|X_j)}$, for each of the values of X_j. The linear model specifying the relationship between Y and X requires that the conditional means of $Y|X$ fall precisely on a straight line defined by the model as illustrated in Figure 1.1 for a single predictor variable. Linear models with two predictors require that the regression surface defined by $\mathbf{X\beta}$ be a two-dimensional plane with partial slopes defining the X axes of the graph as shown in Figure 1.2. For the simple regression model of Equation 1.1, the parameter β_0 defines the expected value of $Y|X = 0$ and β_1 defines the expected rate of change in Y per unit change in X. From the example data of Table 1.1, the regression function of Y = TMT-B on X = Age would appear as in Figure 1.1, in which the conditional means of Y (time to completion of TMT-B) given three values of X = 40, 50, and 75, for example (i.e., $E\{Y_1\}$, $E\{Y_2\}$, $E\{Y_3\}$), lie precisely on the regression line to satisfy the assumption of linearity. Note that the values of the observations Y_1, Y_2, and Y_3 appear in the plane of their respective probability distribution but deviate from their conditional mean. The vector of deviations, $\boldsymbol{\varepsilon} = \mathbf{y} - \mathbf{X\beta}$, are the error terms of the regression model in Equation 1.3.

Figure 1.2 Regression of Trail Making Test–Part B on Age and Education

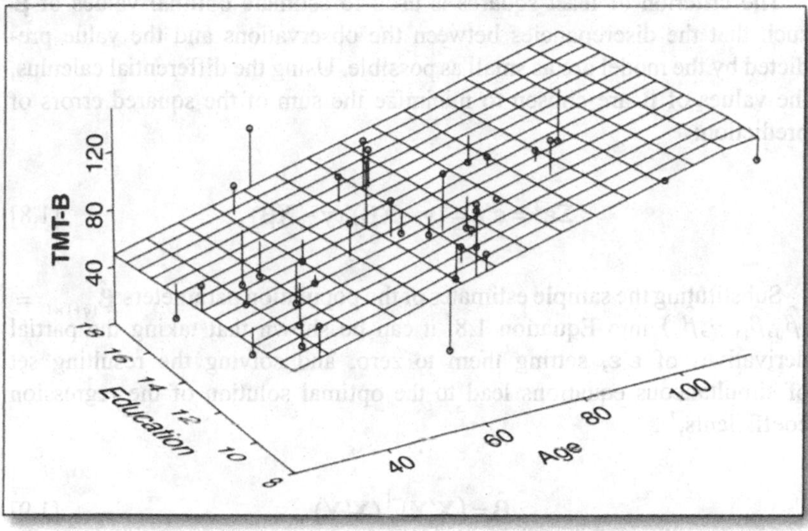

A similar example of a two-predictor model is illustrated in Figure 1.2 by the graph of the relationship between Y = TMT-B, X_1 = age, and X_2 = education for the $n = 40$ sample data descriptively summarized in Table 1.1. The scatterplot reveals a positive relationship between Y and X_1 and a negative relationship between Y and X_2. The population regression function $E(Y|X) = \mathbf{X\beta}$ is defined by the planar surface with partial slopes of β_{X_1} and β_{X2}. The discrepancies between the observations and the model (i.e., the distance between the circles and the plane) are indices of lack of model fit and are captured in the errors of the model, $\boldsymbol{\varepsilon} = \boldsymbol{Y} - \mathbf{X\beta}$.

Thus, all univariate linear models in which the observations are decomposed into model and error components can be written as

$$\mathbf{y} = \mathbf{X\beta} + \boldsymbol{\varepsilon}. \quad [1.6]$$

The differences between Y and the expected values of Y are the errors of prediction of the model,

$$\boldsymbol{\varepsilon} = \mathbf{y} - E(\mathbf{y}|\mathbf{X})^{6}, \quad [1.7]$$

[6]The symbols $\widehat{Y}$, $\widehat{\mathbf{y}}$, $\widehat{\mathbf{Y}}$, and $\widehat{\mu}_{(y|X_1 X_2 \cdots X_q)}$ will denote sample estimates of the population $E(Y|X)$.

which are illustrated by the distance from each point to the two-dimensional plane in Figure 1.2. The closer all the observed values are to the fitted regression plane, the better the fit of the model to the data.

The criterion of least squares is used to estimate optimal values of $\boldsymbol{\beta}$ such that the discrepancies between the observations and the value predicted by the model are as small as possible. Using the differential calculus, the values of $\boldsymbol{\beta}$ are chosen to minimize the sum of the squared errors of prediction:

$$\Sigma\boldsymbol{\varepsilon}^2 = \boldsymbol{\varepsilon}'\boldsymbol{\varepsilon} = (\mathbf{y} - \mathbf{X}\boldsymbol{\beta})'(\mathbf{y} - \mathbf{X}\boldsymbol{\beta}). \qquad [1.8]$$

Substituting the sample estimates of the population parameters $\hat{\boldsymbol{\beta}}_{(q+1\times 1)} = (\hat{\beta}_0, \hat{\beta}_1, \cdots, \hat{\beta}_q)$ into Equation 1.8, it can be shown that taking the partial derivatives of $\boldsymbol{\varepsilon}'\boldsymbol{\varepsilon}$, setting them to zero, and solving the resulting set of simultaneous equations lead to the optimal solution of the regression coefficients,[7]

$$\hat{\boldsymbol{\beta}} = (\mathbf{X}'\mathbf{X})^{-1}(\mathbf{X}'\mathbf{Y}). \qquad [1.9]$$

Applying Equation 1.8 to the example data of Table 1.1 gives the unstandardized parameter estimates of the regression of TMT-B on age, education, and gender,[8]

$$\hat{\boldsymbol{\beta}} = \begin{bmatrix} \hat{\beta}_0 \\ \hat{\beta}_1 \\ \hat{\beta}_2 \\ \hat{\beta}_3 \end{bmatrix} = \begin{bmatrix} 65.69 \\ 0.92 \\ -1.87 \\ -4.68 \end{bmatrix}.$$

[7]We will use the diacritic ^ over the symbol to denote a sample estimate of its population parameter.

[8]$(\mathbf{X}'\mathbf{X})^{-1}$ is the inverse of the uncorrected raw score sum of squares and cross products matrix (SSCP) of **X** and $(\mathbf{X}'\mathbf{Y})$ is the uncorrected raw score sum of cross products (SCP) between **X** and **Y**. The unstandardized regression coefficients of Equation 1.8 are identical to those obtained by mean corrected SSCP and SCP matrices. Details of the relationship between raw score and mean corrected SSCP and SCP matrices are given in Rencher (1998, pp. 269–271).

Interpretations follow the usual rules: Each one year increase in age is accompanied by an increase of approximately $\frac{9}{10}$ of a second to complete the TMT-B task; each additional year of education reduces the time-to-completion of about 2 seconds; and males and females differ by an average of about 4.7 seconds on the timed TMT-B where females show faster performance. The expected time to completion of the TMT-B for a 50-year-old woman with 12 years of education would be estimated at 85 seconds.

It is occasionally useful to reparameterize the regression model to mean zero and unit variance (e.g., $Z_Y, Z_{X_1}, Z_{X_2}, Z_{X3}$) in which the particulars of the regression model in standard score form[9] can be expressed in terms of correlation coefficients. The standard score regression model can be written in scalar and matrix form as

$$Z_Y = \beta_1^* Z_{X_1} + \beta_2^* Z_{X_2} + \cdots + \beta_q^* X_q + \epsilon$$

$$\mathbf{Z_y} = \mathbf{Z_X}\boldsymbol{\beta}^* + \boldsymbol{\epsilon}, \qquad [1.10]$$

with errors of prediction defined as

$$\boldsymbol{\epsilon} = \mathbf{Z_y} - \mathbf{Z_X}\boldsymbol{\beta}^*. \qquad [1.11]$$

The least squares estimates, $\hat{\boldsymbol{\beta}}^*$, of the standardized regression parameters chosen to minimize the sum of squared errors of Equation 1.11 are found by

$$\hat{\boldsymbol{\beta}}^* = \mathbf{R}_{\mathbf{XX}}^{-1}\mathbf{R}_{\mathbf{XY}}, \qquad [1.12]$$

where $\mathbf{R}_{XX}$ and $\mathbf{R}_{XY}$ are, respectively, the correlation matrices between predictors and between predictors and criterion.[10] Estimating $\hat{\boldsymbol{\beta}}^*$ for the example data of Table 1.1 yields the fitted model,

$$\hat{\boldsymbol{\beta}}^* = \begin{bmatrix} \hat{\beta}_1^* \\ \hat{\beta}_2^* \\ \hat{\beta}_3^* \end{bmatrix} = \begin{bmatrix} 0.61 \\ -0.15 \\ -0.07 \end{bmatrix}.$$

[9]The symbol β^* will be used to denote parameters in standard score form with the standardized estimates of the parameters denoted by $\hat{\beta}^*$.

[10] $\mathbf{R}_{XX}$ and $\mathbf{R}_{XY}$ are the sample size–adjusted SSCP and SCP matrices in standard score form.

The usual rules for interpreting standardized coefficients apply; each coefficient represents a $\hat{\beta}_j^*$ standard deviation change in Y per standard deviation change in X_j. There may be little to be gained by interpreting any single standardized regression coefficient in lieu of its unstandardized counterpart, but it is often recommended that standardized coefficients be used if comparative evaluation of the relative influence of predictors is a goal of the analysis (Bring, 1994; Darlington, 1990, pp. 217–218). These recommendations are based on the fact that the absolute value of the unstandardized regression coefficients ($\hat{\beta}_j$) are partly dependent on the scale of measurement, which can differ across predictors while the standardized coefficients ($\hat{\beta}_j^*$) are scale adjusted.[11] For the predictor variables of age and education, the raw regression coefficients suggest that age is a less important predictor than education (ignoring the differences in scale—$SD_{\text{age}} = 21.68, SD_{\text{education}} = 2.60$), whereas the standardized coefficients suggest the opposite relative importance with age being greater than education after adjusting for underlying scale differences. The issue of testing the significance of these differences (i.e., $\hat{\beta}_1$ vs. $\hat{\beta}_2$, and $\hat{\beta}_1^*$ vs. $\hat{\beta}_2^*$) will be shown in a later section to be tests of quite different conceptual hypotheses even if the raw scores are on equal scales, where $SD_{X_1} = SD_{X_2}$.

Assumptions Needed to Justify the Validity of the Least Squares Estimates

There are no assumptions required to justify the least squares estimation of the parameters—that process is purely descriptive. But several important assumptions about the linear model can be introduced at this point. If met, the assumptions provide a degree of confidence in the interpretation of the coefficients as well as justify the validity of the test statistics to be discussed in a later section of this chapter. The assumptions include the following:

- The model is linear; the $E(Y|X)$ lies precisely on a straight line.
- The model is correctly specified; no important variables are omitted from the analysis.

[11]Standardized regression coefficients have little meaning for categorical predictor variables. The standard deviation of the numbers used to designate categories of a nominal grouping variable has no meaningful interpretation beyond the ability of the numerals to distinguish categories. In a later section, we note that the standardized version of a dichotomous predictor may have a useful interpretation when involved in a test of relative importance when compared with other predictors in the model.

- The predictor variables X_j are measured without error.
- $E(\varepsilon)=0$. The errors of the regression model are a random variable with mean zero.
- $Cov(\varepsilon_i,\varepsilon_j)=0$, for i ≠ j. The errors are assumed to be independent with covariance of zero.
- $V(\varepsilon)=\sigma^2 I_{(n\times n)}$. The variance of the errors is assumed to be a constant. The quantity σ^2 is a population parameter and is estimated in the sample by the mean square error,

$$\hat{\sigma}_2 = \frac{\Sigma(Y - X\hat{\mathrm{B}})^2}{n - q_f - 1},$$

  where $n - q_f - 1$ denotes the degrees of freedom for error based on q_f predictor variables in the full model.
- $\varepsilon_i \sim N(X\beta, \sigma^2 I)$. The errors of the model are assumed to be normally distributed with mean $X\beta$ and variance $\sigma^2 I$, which provides the connection to the probability distribution that underlies the test statistics applied to the regression coefficients.

More extensive accounts of the assumptions and the diagnosis of their violations can be found in Cohen et al. (2003, Sect. 4.3–4.5).

Partitioning the Sums of Squares and Defining Measures of Goodness of Fit

The strength of the relationship between the criterion and the predictor variables in a linear model is documented by two indices: the sum of squared errors $\left(SS_{ERROR} = \sum\left(Y - X\hat{\beta}\right)^2 = \sum\hat{\varepsilon}_2 = \hat{\boldsymbol{\varepsilon}}'\hat{\boldsymbol{\varepsilon}}\right)$ and the squared multiple correlation coefficient (R^2). To achieve each of these measures requires that the variability in the response variable be partitioned into its constituent parts related to Equation 1.3. The partitioned *SS* is

$$SS_{TOTAL} = SS_{MODEL} + SS_{ERROR}. \qquad [1.13]$$

The estimated vector of errors of the model is given by $\hat{\boldsymbol{\varepsilon}} = \mathbf{y} - \mathbf{X}\hat{\boldsymbol{\beta}}$ and the sum of squared errors of Equation 1.13 is defined by $\hat{\boldsymbol{\varepsilon}}'\hat{\boldsymbol{\varepsilon}}$. As a measure of goodness of fit, $\hat{\boldsymbol{\varepsilon}}'\hat{\boldsymbol{\varepsilon}}$ has known lower and upper bounds, $0 \le \hat{\boldsymbol{\varepsilon}}'\hat{\boldsymbol{\varepsilon}} \le SS_{TOTAL}$, defining a range from no relationship to perfect relationship. The measure $\hat{\boldsymbol{\varepsilon}}'\hat{\boldsymbol{\varepsilon}}$ is ambiguous as a measure of strength of association unless SS_{TOTAL} is known. The *mean corrected* total sum of squares is $SS_{TOTAL} = \sum\left(Y - \bar{Y}\right)^2 =$

$\mathbf{y'y} - \overline{\mathbf{y}}\overline{\mathbf{y}}n$, where $\overline{\mathbf{y}}$ is an $(n \times 1)$ vector of the mean of *Y* repeated *n* times. Redefining $\mathbf{y'y} = (\mathbf{y'y} - \overline{\mathbf{y}}\overline{\mathbf{y}}n)$ to be the *mean corrected* SS_{TOTAL}, and redefining $\hat{\boldsymbol{\beta}}'\mathbf{X'y} = (\hat{\boldsymbol{\beta}}'\mathbf{X'y} - \overline{\mathbf{y}}\overline{\mathbf{y}}n)$ to represent the *mean corrected* SS_{MODEL}, the partition of the sums of squares of Equation 1.13 is[12]

$$\mathbf{y'y} = \hat{\boldsymbol{\beta}}'\mathbf{X'y} + \boldsymbol{\varepsilon}'\boldsymbol{\varepsilon}. \qquad [1.14]$$

It is common practice to rely on the value of R^2, which is scaled to take on values in the interval [0, 1], as an index of goodness of fit. SS_{TOTAL} is the maximum variability available in *Y*, SS_{ERROR} is the variability in *Y* that cannot be accounted for by the model, and SS_{MODEL} is that part of the variability in *Y* that is accounted for by the model. The proportion of the variability in *Y* that is accounted for by the model, R^2, is the scaled measure of goodness fit and is computed as

$$R^2_{Y \cdot X_1 X_2 \ldots X_q} = 1 - \frac{\boldsymbol{\varepsilon}'\boldsymbol{\varepsilon}}{\mathbf{y'y}} \qquad [1.15]$$

or more commonly,

$$R^2_{Y \cdot X_1 X_2 \ldots X_q} = \frac{\hat{\boldsymbol{\beta}}'\mathbf{X'y}}{\mathbf{y'y}}. \qquad [1.16]$$

If the partitioning is done in terms of standard scores, it can be shown that a convenient definition of R^2 is given by

$$R^2_{Y \cdot X_1 X_2 \cdots X_q} = \hat{\beta}^*_1 r_{Y \cdot X1} + \hat{\beta}^*_2 r_{Y \cdot X2} + \cdots + \hat{\beta}^*_q r_{Y \cdot Xq}. \qquad [1.17]$$

For the TMT-B example data of Table 1.1, the mean corrected Total and Model *SS* are $\mathbf{y'y} = 41875.33$, $\hat{\boldsymbol{\beta}}'\mathbf{X'y} = 17758.00$. The fit of the model is found to be

$$R^2_{Y \cdot X_1 X_2 X_3} = \frac{17758.00}{41875.33} = .424.$$

[12]The uncorrected sum of squares of *Y*, $\sum Y^2 = \mathbf{y'y}$, contains both the *SS* associated with the predictor variables $(\beta_1, \beta_2, \ldots \beta_q)$ and the *SS* associated with the intercept. The mean corrected *SS*, $\mathbf{y'y} - \overline{\mathbf{y}}\overline{\mathbf{y}}n$ disaggregates these two quantities. Rencher (1998, Sect. 4.3–4.5) gives details of the relationships between uncorrected and mean-corrected SS.

About 42% of the variation in TMT-B performance is accounted for by age, education, and gender. About 58% of the variation in Y remains unexplained and is a function of other unknown sources, both systematic and random.

Full and Restricted Models and Squared Semipartial Correlations

In addition to the full model R^2 based on q_f predictors, it is often of interest to ascertain the proportion of variation in Y that is uniquely attributable to X_j adjusted for all remaining X-variables. These *squared semipartial correlation coefficients* $(r^2_{Y(X_1|X_2X_3\cdots X_{q_f})}, r^2_{Y(X_2|X_1X_3\cdots X_{q_f})}, \ldots, r^2_{Y(X_{q_f}|X_1X_2\cdots X_{q_f-1})})$ can be computed from the extra sums of squares approach (Draper & Smith, 1998, pp. 149–160), which requires evaluating the difference between full and restricted model R^2s. Define the *full model* R^2_{full} as the proportion of variation in Y accounted for by all the q_f predictors in the model, $X_1, X_2, \ldots, X_{q_f}$. Define a *restricted model* $R^2_{restricted}$ as the proportion of variability in Y accounted for by a subset of $q_r < q_f$ predictors, say $X_1, X_2, \ldots, X_{q_r}$. Since the full model R^2_{full} documents the proportion of variability in Y accounted for by all the predictors and the restricted model $R^2_{restricted}$ represents the proportion of the variability in Y accounted for by the q_r predictors, the difference between the full and restricted model R^2s must represent the unique incremental variation in Y accounted for by those predictors that are not contained in the restricted model. The difference $R^2_{full} - R^2_{restricted}$ is the squared semipartial correlation coefficient. Examples of squared semipartial correlations for the TMT-B example data are

$$r^2_{Y(X_1|X_2X_3)} = R^2_{Y\cdot X_1X_2X_3} - R^2_{Y\cdot X_2X_3} = .424 - .065 = .359,$$

$$r^2_{Y(X_2|X_1X_3)} = R^2_{Y\cdot X_1X_2X_3} - R^2_{Y\cdot X_1X_3} = .424 - .403 = .021,$$

$$r^2_{Y(X_3|X_1X_2)} = R^2_{Y\cdot X_1X_2X_3} - R^2_{Y\cdot X_1X_2} = .424 - .419 = .005,$$

$$R^2_{Y(X_1X_2|X_3)} = R^2_{Y\cdot X_1X_2X_3} - R^2_{Y\cdot X_3} = .424 - .002 = .422.$$

About 36% of the variation in TMT-B performance is attributable to age after adjusting for the variance accounted for by education and gender; the variance in TMT-B that is uniquely attributable to education or gender is negligible. The multiple squared semipartial of age and education adjusted for gender appears to be the best prediction model, but it is unclear if this value is a significant improvement over age alone $(r^2_{YX_1} = .6324^2 = .400)$ because we know little about the sampling variability that accompanies these models. Methods for assessing statistical significance and testing hypotheses on contrasts between predictors are reviewed in the next section.

Testing Hypotheses on the Regression Coefficients and R^2s

The trustworthiness of $\hat{\beta}$ or R^2 depends on knowledge of the sampling variability of that statistic and test statistics for evaluating hypotheses on the parameters of the model. The two most common methods include the F-test on values of R^2 and the single degree of freedom t-test on the model regression coefficient where $t = \sqrt{F}$. A generic F-test on df_h and df_e degrees of freedom based on appropriately specified full and restricted models can be defined as

$$F_{(df_h,\, df_e)} = \frac{R^2_{full} - R^2_{restricted}}{1 - R^2_{full}} \cdot \frac{df_e}{df_h}. \qquad [1.18]$$

Let q_f be the number of predictors in the full model (exclusive of the unit vector X_0), let q_r be the number of predictors in the restricted model, and let $df_h = q_f - q_r$ and $df_e = n - q_f - 1$. If it is assumed that $\varepsilon_i \sim N(0, \sigma^2)$, then the test statistic in Equation 1.18 follows the F distribution with $q_f - q_r$ and $n - q_f - 1$ degrees of freedom. The numerator of the left-most ratio of the F-test is the definition of the squared semipartial correlation. The nature of $R^2_{restricted}$ will be dictated by the hypothesis to be tested since the hypothesis dictates the constraints to be placed on the full model. If the hypothesis on the whole model $H_0 : \beta_1 = \beta_2 = \cdots = \beta_{q_f} = 0$ is desired,[13] the restricted model will contain only β_0 with $R^2_{restricted} = 0$, leading to the numerator $R^2_{full} - R^2_{restricted} = R^2_{full}$. A test of a hypothesis on a single regression coefficient, such as $H_0 : \beta_1 = 0$, would involve $R^2_{restricted} = R^2_{Y \cdot X_2 X_3 \cdots X_{q_f}}$. Hypotheses on any single coefficient, or set of coefficients, can be tested in this manner. Further examples of hypothesis tests involving restrictions placed on the linear model are given in Rindskopf (1984).

For single df_h tests, the t-test on the hypothesis $H_0 : \beta_j = k$ is, in common usage,[14]

$$t_{(df_e)} = \frac{\hat{\beta}_j - \beta_j}{\sqrt{\dfrac{MSE}{SS_{X_j}} \left(\dfrac{1}{1 - R^2_{X_j \cdot other}} \right)}}, \qquad [1.19]$$

---

[13]This is equivalent to the hypothesis $H_0 : \rho^2_{Y \cdot X_1 X_2 \cdots X_{q_f}} = 0$.

[14]The value of k need not be hypothesized to be 0; any theoretically defensible value of k is permissible.

where

$$MSE = \frac{SS_{ERROR}}{n - q_f - 1},$$

SS_{x_j} is the sum of squares of the predictor variable involved in the test, and

$$\frac{1}{1 - R^2_{X_j \cdot other}}$$

is the variance inflation factor (VIF) that adjusts for the multicollinearity among the predictor variables. For the TMT-B example, the F-test on the whole model $R^2 = .424$ is $F_{(3,36)} = 8.84, p < .001$. The test of the significance of each of the individual partial regression coefficients for age, education, and gender yielded, respectively, values of $t_{(36)} = 4.74, p < .001$; $t_{(36)} = -1.15, p = .258$; and $t_{(36)} = .57, p = .575$. Only the age variable is uniquely related to TMT-B performance. The t-test statistics on the individual coefficients are the $\sqrt{F}$ that would have been obtained by the full- versus restricted-model approach of Equation 1.18. The results of the test of hypotheses on the values of β_j and on the values of their respective partial and semipartial correlations are identical.

The General Linear Hypothesis Test

Although the two methods for testing hypotheses described above are in wide usage, they are special cases of a much more general approach to testing hypotheses in linear models—the general linear hypothesis test. The general linear test is a compact procedure that covers an astonishing array of common and specialized tests of hypotheses in both the univariate and the multivariate linear models.

Assume for the univariate model of Equation 1.3 that we wish to test the hypothesis that all the sample regression coefficients in a full model have been drawn from a population in which all the coefficients, with the exception of the intercept, are simultaneously equal to zero. We can formalize this hypothesis by a linear combination of the parameters specified by the matrix product $\mathbf{L\beta} = \mathbf{0}$. That is,

$$H_0: \mathbf{L\beta} = \begin{bmatrix} 0 & 1 & 0 & \cdots & 0 \\ 0 & 0 & 1 & \cdots & 0 \\ \vdots & \vdots & \vdots & \cdots & \vdots \\ 0 & 0 & 0 & \cdots & 1 \end{bmatrix} \begin{bmatrix} \beta_0 \\ \beta_1 \\ \beta_2 \\ \vdots \\ \beta_{q_f} \end{bmatrix} = \begin{bmatrix} \beta_1 \\ \beta_2 \\ \vdots \\ \beta_{q_f} \end{bmatrix} = \begin{bmatrix} 0 \\ 0 \\ \vdots \\ 0 \end{bmatrix}. \qquad [1.20]$$

The **L** matrix is of order ($c \times q_f + 1$) whose role is to identify the coefficients of interest in any hypothesis. Other hypotheses might involve only a single-parameter estimate (e.g., $H_0: \beta_1 = 0$), or some subset of the parameters $\left(\text{e.g., } H_0: \begin{bmatrix} \beta_1 \\ \beta_3 \end{bmatrix} = \begin{bmatrix} 0 \\ 0 \end{bmatrix} \right)$. In general, any desired hypothesis can be defined as a product of a vector (or matrix) of *contrast coefficients*, $\mathbf{L}_{(c \times q+1)}$, and the vector of parameters, $\boldsymbol{\beta}_{(q+1 \times 1)}$, from the full model analysis. A more general form of the contrast is possible where the vector **k** can contain zeros (the traditional null hypothesis) or any other vector of theoretically justified nonzero values:

$$\mathbf{L}_{(c \times q+1)} \boldsymbol{\beta}_{(q+1 \times 1)} = \mathbf{k}_{(c \times 1)}. \qquad [1.21]$$

The subscript c denotes the number of rows in **L** that will be equivalent to df_h in the associated test statistic. Once the desired hypothesis is specified, we can substitute the estimates of the parameters $\hat{\boldsymbol{\beta}}$ into Equation 1.22 to obtain the sums of squares for the hypothesis:

$$SS_{HYPOTHESIS} = (\mathbf{L}\hat{\boldsymbol{\beta}})' \left(\mathbf{L} (\mathbf{X}'\mathbf{X})^{-1} \mathbf{L}' \right)^{-1} (\mathbf{L}\hat{\boldsymbol{\beta}}) \qquad [1.22]$$

and the $SS_{HYPOTHESIS}$ can be used as the numerator of a familiar version of the F-test,

$$F_{(df_h,\, df_e)} = \frac{SS_{HYPOTHESIS}}{SS_{ERROR}} \bullet \frac{df_e}{df_h}. \qquad [1.23]$$

Under the assumption that the errors of the model are normally distributed, F will follow the F distribution on $df_h = c$ and $df_e = n - q_f - 1$ degrees of freedom.

The Test of the Whole Model Hypothesis
$\beta_1 = \beta_2 = \beta_3 = 0$ and $\rho^2_{Y \cdot X_1 X_2 X_3}$

For the TMT-B example data, we found the estimated regression coefficients to be

$$\hat{\boldsymbol{\beta}} = \begin{bmatrix} \hat{\beta}_0 \\ \hat{\beta}_1 \\ \hat{\beta}_2 \\ \hat{\beta}_3 \end{bmatrix} = \begin{bmatrix} 65.69 \\ 0.92 \\ -1.87 \\ -4.68 \end{bmatrix}.$$

and we desire a test the hypothesis that parameters for X_1, X_2, and X_3 are simultaneously equal to 0, $H_0: \beta_1 = \beta_2 = \beta_3 = 0$. This statement is also a test of $H_0: \rho^2_{Y \cdot X_1 X_2 X_3} = 0$. The general linear test of the full model hypothesis is given in $\mathbf{L\beta}$

$$\mathbf{L\beta} = \begin{bmatrix} 0 & 1 & 0 & 0 \\ 0 & 0 & 1 & 0 \\ 0 & 0 & 0 & 1 \end{bmatrix} \begin{bmatrix} \beta_0 \\ \beta_1 \\ \beta_2 \\ \beta_3 \end{bmatrix} = \begin{bmatrix} \beta_1 \\ \beta_2 \\ \beta_3 \end{bmatrix} = \begin{bmatrix} 0 \\ 0 \\ 0 \end{bmatrix},$$

which ignores the intercept. For the contrast matrix $\mathbf{L}$, the inverse of the sum of squares and cross-products matrix among the three predictor variables, $\mathbf{X'X}^{-1}$, and the estimates of the parameters $\hat{\boldsymbol{\beta}}$ the hypothesis SS of Equation 1.22 is

$$SS_{HYPOTHESIS} = \left(\begin{bmatrix} 0 & 1 & 0 & 0 \\ 0 & 0 & 1 & 0 \\ 0 & 0 & 0 & 1 \end{bmatrix} \begin{bmatrix} 65.69 \\ 0.92 \\ -1.87 \\ -4.68 \end{bmatrix} \right)'$$

$$\left(\begin{bmatrix} 0 & 1 & 0 & 0 \\ 0 & 0 & 1 & 0 \\ 0 & 0 & 0 & 1 \end{bmatrix} \begin{bmatrix} 40 & 2{,}339 & 504 & 18 \\ 2{,}339 & 155{,}103 & 29{,}097 & 1{,}059 \\ 504 & 29{,}097 & 6{,}614 & 221 \\ 18 & 1059 & 221 & 18 \end{bmatrix}^{-1} \begin{bmatrix} 0 & 0 & 0 \\ 1 & 0 & 0 \\ 0 & 1 & 0 \\ 0 & 0 & 1 \end{bmatrix} \right)^{-1}$$

$$\left(\begin{bmatrix} 0 & 1 & 0 & 0 \\ 0 & 0 & 1 & 0 \\ 0 & 0 & 0 & 1 \end{bmatrix} \begin{bmatrix} 65.69 \\ 0.92 \\ -1.87 \\ -4.68 \end{bmatrix} \right)$$

which yields $SS_{HYPOTHESIS} = 17758.00$. The $SS_{HYPOTHESIS}$ is identical to the SS_{MODEL} obtained from $\hat{\boldsymbol{\beta}}'\mathbf{X}'\mathbf{y} - \bar{\mathbf{y}}'\bar{\mathbf{y}}n$. With $df_h = c = 3$, $df_e = n - q_f - 1 = 36$, and $\hat{\boldsymbol{\varepsilon}}'\hat{\boldsymbol{\varepsilon}} = 24117.33$, the F-test on the whole model association is found to be

$$F_{(3,16)} = \frac{17758}{24117.33} \cdot \frac{36}{3} = 8.84, p = .0002.$$

With $R^2_{Y \cdot X_1 X_2 X_3} = .424$, there is sufficient evidence to reject H_0.

Testing the Individual Contributions of X_1, X_2, and X_3 by the General Linear Test

Hypothesis tests on the individual partial regression coefficients β_1,β_2, and β_3 can be readily tested within the $\mathbf{L\beta=0}$ framework. For testing a hypothesis on β_1, we specify

$$\mathbf{L\beta}=\begin{bmatrix}0 & 1 & 0 & 0\end{bmatrix}\begin{bmatrix}\beta_0\\ \beta_1\\ \beta_2\\ \beta_3\end{bmatrix}=\beta_1=0 \qquad [1.24]$$

and to specify the null hypothesis on β_2 and on β_3 we employ the vector $\boldsymbol{\beta}$ and the appropriate vectors $\mathbf{L}=[0 \quad 0 \quad 1 \quad 0]$ and $\mathbf{L}=[0 \quad 0 \quad 0 \quad 1]$, respectively. All these hypotheses are tested by substituting $\hat{\boldsymbol{\beta}}$ into Equations 1.22 and 1.23; the results are summarized in Table 1.2.

Table 1.2 General Linear Hypothesis Tests on Individual Partial Regression Coefficients

Hypothesis	$\hat{\beta}$	$\hat{\beta}^*$	$r_{semipartial}$	$F_{(1,16)}$	p
Age: $\beta_1 = 0$	0.919	0.608	.599	22.44	<.001
Education: $\beta_2 = 0$	−1.870	−0.148	−.145	1.32	.258
Gender: $\beta_3 = 0$	-4.683	−0.072	−.072	0.32	.575

In this model, age is the only significant contributor to the prediction of TMT-B. The test statistic on the any unstandardized $\hat{\beta}_j$ is also the test of the significance of the standardized $\hat{\beta}^*$ and the semipartial correlation $r_{Y(X_j|X_1X_2\cdots)}$. The test of hypotheses on sets of predictors is also identical for unstandardized and standardized partial regression coefficients and the multiple semipartial correlations associated with each set. These equivalences no longer hold when more complex hypotheses are tested by the general linear test.

Testing More Complex Hypotheses With the General Linear Test

The general linear hypothesis test is suitable for formulating and testing many complex hypotheses (Draper & Smith, 1998, pp. 217–221;

Rindskopf, 1984). Consider the question of whether age is a *better* predictor of TMT-B performance than is education, both adjusted for gender. This question asks if the coefficients $\hat{\beta}_1$ and $\hat{\beta}_2$ are *significantly different from one another* as expressed in the null hypothesis $H_0: \beta_1 = \beta_2$. This hypothesis can be specified by the contrast matrix $\mathbf{L} = [0 \quad 1 \quad -1 \quad 0]$ that deletes β_0 and β_3 from $\boldsymbol{\beta}$ and defines the difference between β_1 and β_2. The symbolic contrast of the null hypothesis, $\mathbf{L\beta} = \mathbf{0}$ is

$$\mathbf{L\beta} = [0 \quad 1 \quad -1 \quad 0]\begin{bmatrix}\beta_0 \\ \beta_1 \\ \beta_2 \\ \beta_3\end{bmatrix} = [\beta_1 - \beta_2] = 0, \qquad [1.25]$$

which gives the basis for evaluating the $SS_{HYPOTHESIS}$ and the numerator of the F-test. Substituting the estimates $\hat{\beta}_j$ into Equation 1.22 we find,

$$\mathbf{L\hat{\beta}} = [0 \quad 1 \quad -1 \quad 0]\begin{bmatrix}65.69 \\ .92 \\ 1.87 \\ -4.68\end{bmatrix} = [-.95]$$

and $\left(\mathbf{L}(\mathbf{X'X})^{-1}\mathbf{L'}\right)^{-1} = 259.53$ with $\hat{\boldsymbol{\varepsilon}}'\hat{\boldsymbol{\varepsilon}} = 24117.33$. The F-test of Equation 1.23 is

$$F_{(1,16)} = \frac{234.71}{24117.33} \cdot \frac{36}{1} = 0.32, p = .573.$$

The unstandardized partial slopes of age and education (reverse scored),[15] shown in Figure 1.3, do not differ significantly from one

---

[15]Age has a positive relationship to TMT-B; performance deteriorates with increasing age. Conversely, TMT-B has a negative relationship with increasing years of education. Contrasts between regression coefficients are sensitive to both magnitude and direction and a choice must be made between testing differences in magnitude only, or testing differences in both magnitude and direction. Theoretical considerations based on substance knowledge should be brought to bear to make this choice. For the age versus education comparison illustrated here, only the magnitude of the effect is of interest. Reversing the scoring of the education variable equates the sign of both age and education coefficients; hence the contrast is one of magnitude and not direction. If there is theoretical justification to leave the signs of the regression coefficients in the original scoring of age and education, then a test of both magnitude and direction would result. The F-test on this contrast is $F_{(1,16)} = 3.01$, $p = .091$, still a nonsignificant result.

another. Interpreting this lack of a significant difference should be done cautiously. Many authors point out that such a contrast makes sense only if the two variables are measured on the same scale, which is not the case with age (range = 33–105, *SD* = 21.7) and education (range = 8–18, *SD* = 2.6).

The unstandardized partial slopes of Figure 1.3 do not differ significantly from one another, but the squared semipartial correlations suggest that the variance in TMT-B accounted for by age is substantially greater than the variance accounted for by education. The relative rank order of the two predictor variables is opposite when unstandardized slopes and semipartial correlations are used for the ranking, largely due to the differences of scale of the predictor variables.

An alternative test that avoids the issue of inequalities of scale is the general linear test applied to standardized coefficients by testing the hypothesis $H_0 : \beta_1^* - \beta_2^* = 0$.[16] Estimating the parameters by $\hat{\beta}_1^*$ and $\hat{\beta}_2^*$ and

Figure 1.3 Comparison of Unstandardized Partial Slopes, Standardized Partial Slopes, and Semipartial Correlations

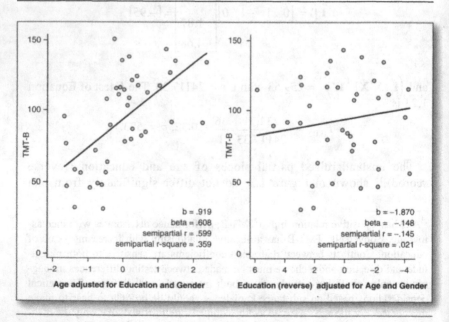

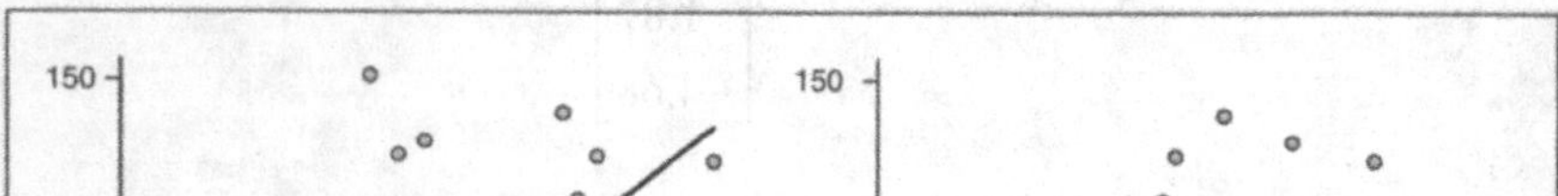

Note: Education is reverse scored to guarantee a positive slope.

[16]The scoring of the education variable was also reversed in this analysis to constrain the sign of each standardized slope to a positive value. The contrast is therefore a test of the difference in magnitude of semipartial correlations.

performing the same sequence of computations on the standardized variables Z_Y, Z_{X_1}, Z_{X_2}, and Z_{X_3} leads to

$$\mathbf{L}\hat{\boldsymbol{\beta}}^* = [1 \quad -1 \quad 0]\begin{bmatrix} 0.61 \\ 0.15 \\ -0.07 \end{bmatrix} = .460,$$

$SS_{HYPOTHESIS} = 3.397$, $SS_{ERROR} = 22.461$,[17] and $F_{(1,36)} = 5.44, p = .025$. The standardized parameter estimates differ significantly by the hypothesis test applied to standardized coefficients. The reason for the differing results is a consequence of the differences in the scales of measurement of the predictor variables; it can be shown that the *j*th standardized coefficient is a ratio of its semipartial correlation to the square root of the proportion of variation in X_j not accounted for by the remaining predictors $X_{j'}$ (e.g., tolerance) in the full model, that is,

$$\hat{\beta}^*_j = \frac{r_{Y(X_j|X_{j'})}}{\sqrt{1-R^2_{j.other}}}.$$

Unstandardized regression coefficients and their standard errors have absolute magnitude for two reasons: (1) scale of measurement and (2) the underlying relationship between the predictor and response variables. Conversely, the magnitude of the standardized coefficients is most heavily determined by the semipartial correlations and the tolerances of the predictors. Consequently, a difference between standardized coefficients constitutes a test of the differences between semipartial correlation coefficients[18]—it is a test of differences between correlations of *Y* and each predictor after adjustment for other predictors in the model. The test statistic of differences between raw regression coefficients and between semipartial correlations need not be equal. The two tests are numerically independent because they test conceptually different hypotheses—differences in rates of change versus differences in strength of association. The tests between coefficients for

---

[17]The error sum of squares in standard score form is $\mathbf{Z}'_Y\mathbf{Z}_Y - \hat{\mathbf{B}}^{*\prime}\,\mathbf{Z}'_X\mathbf{Z}_Y = (n-1)(1-R^2_{Y \cdot X_1X_2X_3})$.

[18]The test of the differences between two standardized regression coefficients from a regression analysis is defined as

$$t = \frac{\hat{\beta}^*_1 - \hat{\beta}^*_2}{\sqrt{MSE\left(\mathbf{L}\mathbf{R}^{-1}_{XX}\mathbf{L}'\right)}}$$

unstandardized and standardized models will be identical only when $S_{X_1} = S_{X_2}$. Similar tests of differences between correlation coefficients are discussed in Olkin and Finn (1995). Draper and Smith (1998, pp. 218–219) and Rencher (1998, pp. 295–300) give examples of more complicated linear hypothesis tests in which the same principles apply.

Generalizing From Univariate to Multivariate General Linear Models

We have begun this volume with a review of the common strategies for modeling a single response variable as a function of one or more continuous and/or categorical explanatory variables. Such models have great flexibility and can accommodate any combination of predictor variable types, including their interactions and powers.

(Cohen et al., 2003, pp. 640–642), where $\mathbf{R}_{XX}^{-1}$ is the inverse of the correlation matrix among the predictors and

$$MSE = \frac{1 - R^2_{Y \cdot X_1 X_2 X_3}}{n - q_f - 1}.$$

Substituting the definitions

$$\hat{\beta}_1^* = \frac{r_{Y(X_1|X_2)}}{\sqrt{1 - R^2_{1.2}}}$$

and

$$\hat{\beta}_2^* = \frac{r_{Y(X_2|X_1)}}{\sqrt{1 - R^2_{1.2}}}$$

into t sets the numerator to

$$\frac{r_{Y(X_1|X_2)} - r_{Y(X_2|X_1)}}{\sqrt{1 - R^2_{1.2}}}.$$

Setting the contrast matrix to $\mathbf{L} = [1 \quad -1 \quad 0]$ and performing the symbolic multiplication of the quantity $\sqrt{MSE\ (\mathbf{L}\mathbf{R}_{XX}^{-1}\mathbf{L}')}$, the denominator of t reduces to

$$\frac{\sqrt{MSE\, 2(1 + r_{12})}}{\sqrt{1 - R^2_{1.2}}}.$$

Recounting these details here sets the stage for the generalization of these same analytic concepts to those instances where *more than one* dependent variable is to be analyzed simultaneously. Models with $p > 1$ response variables are classified as multivariate models that can be treated with the same four-step process—the specification of the multivariate model, estimation of its parameters, identifying measures of strength association, and defining appropriate tests of significance. We pursue these topics in the chapters that follow.

The quantities $\sqrt{1-R_{1.2}^2}$ in the numerator and denominator cancel, leaving

$$t = \frac{r_{Y(X_1|X_2)} - r_{Y(X_2|X_1)}}{\sqrt{MSE\left(2(1+r_{12})\right)}}.$$

Hence, the test of the hypothesis $\beta_1^* - \beta_2^* = 0$ is a test of the differences between semipartial correlation coefficients. In this interpretation, approximately 36% of the variance in TMT-B is accounted for by age while about 2% of the variance in TMT-B is accounted for by education. The absolute values of the two correlations are significantly different from one another, while the absolute values of the two unstandardized slopes do not differ significantly. The difference between unstandardized rates of change is being masked by differences in variance of the predictors.

CHAPTER 2. SPECIFYING THE STRUCTURE OF MULTIVARIATE GENERAL LINEAR MODELS

The transition from the scalar version of the univariate linear model to the univariate model expressed in matrix algebraic terms is given in Chapter 1 (see Equations 1.2 and 1.3). The univariate linear model is readily generalized to the multivariate model with $p > 1$ response variables by augmenting the orders of $\mathbf{Y}, \mathbf{B}$, and $\mathbf{E}$ to accommodate the additional columns of dependent variables, the added columns of regression coefficients associated with each dependent variable, and the additional columns of the disturbances associated with each Y variable in the matrix of errors. To specify the multivariate model, we write

$$\mathbf{Y}_{(n \times p)} = \mathbf{X}_{(n \times q+1)} \mathbf{B}_{(q+1 \times p)} + \mathbf{E}_{(n \times p)}. \qquad [2.1]$$

In this chapter, we will define the elements of these matrices and discuss both the statistical and the substantive ideas necessary to specify the multivariate model that must accommodate multiple columns of $\mathbf{Y}$, $\mathbf{B}$ and $\mathbf{E}$. The order of these three matrices is one key feature of the specification that differentiates multivariate from univariate models. Conversely, the design matrix $\mathbf{X}$, in all of its possible variations, will be identical to the comparable univariate design matrix—we need only develop the mechanism for coping with the multiplicities of dependent variables, parameter estimates, and disturbances that characterize multivariate linear models.

Specifying the multivariate linear model involves at least two discrete, but related, activities:

- Choosing reliable and valid criterion and predictor variables based on theoretical explanations of their hypothesized relationships, including their direction, magnitude, and conceptual mechanism (see Jaccard & Jacoby, 2010, for a discussion of building conceptual theoretical models), and
- Specifying the mathematical model that is consistent with these theoretical arguments.

In this chapter, we introduce methods for specifying the mathematical form of the multivariate model, discuss several different specifications of the design matrix $\mathbf{X}$, and introduce the numerical examples that will be used to illustrate these developments in subsequent chapters.

The Mathematical Specification of the Model

The mathematical specification of the multivariate linear model begins with the definitions of the four matrices of Equation 2.1 that denote a truly multivariate problem if the number of criterion variables (p) is greater than 1. Multivariate models are written in matrix terms, and following the usual conventions[1] we denote the *order* of the matrix by designating its dimensions by reference to the number of rows and columns in the matrix. The intersection of any row and any column defines a specific *element* of the matrix; Y_{23}, for example, denotes the observation of the second row and the third column of $\mathbf{Y}$. Letting n denote the number of observations and p denote the number of dependent variables in a model, then the $(n \times p)$ dependent variable matrix $\mathbf{Y}_{(n \times p)}$ denotes a matrix of n rows and p columns. An expanded version of all such $\mathbf{Y}$ matrices will therefore have a similar general form in which the order of $\mathbf{Y}$ and all of its elements can be readily identified,

$$\mathbf{Y}_{(n \times p)} = \begin{bmatrix} Y_{11} & Y_{12} & \cdots & Y_{1p} \\ Y_{12} & Y_{22} & \cdots & Y_{2p} \\ \vdots & \vdots & \ddots & \vdots \\ Y_{n1} & Y_{n2} & \cdots & Y_{np} \end{bmatrix}.$$

Similarly, the explanatory variables of the model are contained in the *design matrix*, $\mathbf{X}_{(n \times q+1)}$ in which the order of the matrix is defined by the n rows and the $q+1$ column vectors consisting of the q predictor measures $(X_1, X_2, \ldots X_q)$ and the unit column vector $X_0 \equiv 1$ as previously defined for estimating the model intercept. The design matrix will have a general form of

$$\mathbf{X}_{(n \times q+1)} = \begin{bmatrix} 1 & X_{11} & X_{12} & \cdots & X_{1q} \\ 1 & X_{12} & X_{22} & \cdots & X_{2q} \\ \vdots & \vdots & \vdots & \ddots & \vdots \\ 1 & X_{n1} & X_{n2} & \cdots & X_{nq} \end{bmatrix}.$$

The matrix of model parameters $\mathbf{B}$ of Equation 2.1 differs substantially from the univariate model of Equation 1.3. Multiple dependent variables are accompanied by multiple columns of $\mathbf{B}$ to accommodate all of the *Y-X* relationships. The order of $\mathbf{B}$ is governed by the $q+1$ columns of $\mathbf{X}$ and the

---

[1]We assume some familiarity with matrix terminology and matrix algebraic procedures. Detailed coverage is given in Namboodiri (1984) and Schott (1997); succinct coverage relevant to regression analysis is given in Draper and Smith (1998, Chap. 4).

p columns of $\mathbf{Y}$; $\mathbf{B}_{(q+1\times p)}$ defines the matrix of parameters in the population model that must be estimated as part of the analysis. The rows of $\mathbf{B}$ correspond to the predictor variables $X_0, X_1, X_2, \ldots X_q$ and the columns represent the response variables $Y_1, Y_2, \ldots Y_p$,

$$\mathbf{B}_{(q+1\times p)} = \begin{bmatrix} \beta_{01} & \beta_{02} & \cdots & \beta_{0p} \\ \beta_{11} & \beta_{12} & \cdots & \beta_{1p} \\ \beta_{21} & \beta_{22} & \cdots & \beta_{2p} \\ \vdots & \vdots & \ddots & \vdots \\ \beta_{q1} & \beta_{q2} & \cdots & \beta_{qp} \end{bmatrix}.$$

In Equation 2.1, the matrix product $\mathbf{X}_{(n\times q+1)}\mathbf{B}_{(q+1\times p)}$ conforms with respect to multiplication, and the order of the product $\mathbf{XB}_{(n\times p)}$ is determined by the number of rows of $\mathbf{X}$ and the number of columns of $\mathbf{B}$. Following the row-by-column rules for matrix multiplication results in a product matrix that contains the weighted linear combinations of $\mathbf{XB}$ for each of the variables in $\mathbf{Y}$ across all of the n observations:

$$\mathbf{XB}_{(n\times p)} = \begin{bmatrix} 1 & X_{11} & X_{12} & \cdots & X_{1q} \\ 1 & X_{12} & X_{22} & \cdots & X_{2q} \\ \vdots & \vdots & \vdots & \ddots & \vdots \\ 1 & X_{n1} & X_{n2} & \cdots & X_{nq} \end{bmatrix} \begin{bmatrix} \beta_{01} & \beta_{02} & \cdots & \beta_{0p} \\ \beta_{11} & \beta_{12} & \cdots & \beta_{1p} \\ \beta_{21} & \beta_{22} & \cdots & \beta_{2p} \\ \vdots & \vdots & \ddots & \vdots \\ \beta_{q1} & \beta_{q2} & \cdots & \beta_{qp} \end{bmatrix}$$

The additive equality expressed in Equation 2.1 is satisfied since the order of $\mathbf{XB}_{(n\times p)}$ conforms to the order of $\mathbf{E}_{(n\times p)}$, which in turn conforms to the order of $\mathbf{Y}_{(n\times p)}$. Using these results, the expanded matrix version of the full multivariate linear model for the variables $\mathbf{Y}$, $\mathbf{X}$, $\mathbf{B}$, and $\mathbf{E}$ would appear as

$$\begin{bmatrix} Y_{11} & Y_{12} & \cdots & X_{1p} \\ Y_{21} & Y_{22} & \cdots & Y_{2p} \\ \vdots & \vdots & \ddots & \vdots \\ Y_{n1} & Y_{n2} & \cdots & Y_{np} \end{bmatrix} = \begin{bmatrix} 1 & X_{11} & X_{12} & \cdots & X_{1q} \\ 1 & X_{21} & X_{22} & \cdots & X_{2q} \\ \vdots & \vdots & \vdots & \ddots & \vdots \\ 1 & X_{n1} & X_{n2} & \cdots & X_{nq} \end{bmatrix} \begin{bmatrix} \beta_{01} & \beta_{01} & \cdots & \beta_{0p} \\ \beta_{11} & \beta_{01} & \cdots & \beta_{1p} \\ \beta_{21} & \beta_{01} & \cdots & \beta_{3p} \\ \vdots & \vdots & \ddots & \vdots \\ \beta_{q1} & \beta_{01} & \cdots & \beta_{qp} \end{bmatrix}$$

$$+ \begin{bmatrix} \varepsilon_{11} & \varepsilon_{12} & \cdots & \varepsilon_{1p} \\ \varepsilon_{21} & \varepsilon_{22} & \cdots & \varepsilon_{2p} \\ \vdots & \vdots & \ddots & \vdots \\ \varepsilon_{n1} & \varepsilon_{n2} & \cdots & \varepsilon_{np} \end{bmatrix}$$

All the multivariate models to be covered in this volume will be specified by mathematical models consistent with Equation 2.1. The number of units of observation (cases, participants) and the number of the variables in the response matrix $\mathbf{Y}_{(n \times p)}$ and design matrix $\mathbf{X}_{(n \times q+1)}$ determine the initial specification of the model. The remaining aspects of the model specification rest on theoretical and conceptual arguments and will also depend on design considerations (e.g., multivariate multiple regression [MMR] or multivariate analysis of variance [MANOVA]) that will dictate the nature of the vectors of the design matrix $\mathbf{X}_{(n \times q+1)}$.

Defining the Substantive Roles of Criterion and Predictor Variables

The specification of the model in multivariate analysis is partly nonmathematical, and it is best that there be clear reasons and careful definitions for inclusion of both dependent and explanatory variables. Theoretical considerations are paramount in this endeavor. Since theoretical arguments are project specific, we attempt to lay out briefly the conceptual arguments that underlie each of the examples introduced at the end of this chapter. More extensive advice on this important aspect of model specification is given in Jaccard and Jacoby (2010). Beyond the theoretical and conceptual arguments that dictate the choice of response and explanatory variables, there are four general considerations and decision points that apply across all projects that require attention prior to data collection and analysis. They include the following:

- Measurement level of the Y variables
- Measurement of the X-variables; either continuously distributed, categorical, or both
- Experimental status of the X-variables; either manipulated or observed
- Purpose of the X-variables; theoretical substance or control of confounding

The first consideration is the nature of the dependent variables. In this volume, we deal exclusively with continuously distributed dependent variables.[2] Most traditional multivariate analyses have been developed around interval data that can be assumed to be multivariate normal in distribution. Although multivariate models that deal with limited dependent variables such as rank

[2]We refer here to truly continuous variables (an infinite number of possible gradations on the real number line) and discrete variables (a quantitative scale whose integer values are ordinal but on which the gradations between integers is suspect; i.e., 2.5 children). In this volume, we follow the looser tradition of treating both variables as continuous. The difference will be evident by the context of the examples.

transformation analysis (Puri & Sen, 1971), multivariate logistic regression (Glonek & McCullagh, 1995), and cross-classified frequency counts (Zwick & Cramer, 1986) have been proposed, we do not cover them here.

On the predictor variable side of the model several features of the *X*-variables must be considered. These decisions determine how the design matrix will be formulated and how the data are (or have been) collected, how inferences are made from the analysis, and what inferences are justified. The first of these decision points is to decide if the *X*-variable is continuously distributed, discrete, or categorical in nature.[3] A model containing only continuous or discrete explanatory variables is typically classified as a traditional regression model while those models containing only categorical predictor variables are often classified as analysis of variance models. Models with both continuous and categorical predictors in the design matrix have no special designation but are equally possible in the linear model analysis. We present example data sets below that contain continuous explanatory variables, categorical variables (requiring one or more vectors), and combinations of both types of variables.

The second decision that must be considered about the predictor variables is their intended role in inference: Are they theoretically important and require tests of hypotheses, or are they to be treated as covariates for purposes of controlling extraneous variance and potential confounding? A variable's role will usually be clear from a carefully argued theoretical context and is part of the process of specifying the model. The same is true for control variables—their inclusion is based on whether they serve one of two purposes—either they are (1) included because they are known to be substantially correlated with the dependent variables, but not to the theoretically important predictors in the model, such variables included in the design matrix **X** can reduce error variance, or (2) they are substantially correlated with both an explanatory variable and one or more response variables and therefore are serious candidates for common-cause, third-variable confounders (Rothman, Greenland, & Lash, 2008). In both instances, their inclusion is intended to be one of control and may or may not require a hypothesis test on the variable.

---

[3]It is necessary to keep in mind the distinction between a variable (say *X*) and a vector (say $\mathbf{x}_1$). Continuously distributed variables require only a single vector to represent their variability. Categorical variables, such as group membership in multiple groups, require multiple vectors to represent their variability. The *variable* of "treatment" that compares two different treatments with a single control contains three groups and requires two *vectors* to fully represent its variability. Discussions of categorical or qualitative variable coding schemes in linear model analysis can be found in Cohen et al. (2003, Chap. 8).

A final judgment that must be made in the selection of predictor variables in a linear model is related to the experimental versus observational origin of the *X*-variables in the model, namely, what is the underlying source of the variability in the predictor? Is the variability of the explanatory variable under the control of the experimenter or does its variability derive from unknown sources? The first of these sources of variability characterizes the manipulated experiment and the second describes ex post facto observational studies. While it matters little to the mathematical specification of the model, this characteristic of the specification plays an important role in the permissible conclusions that can be drawn from the analysis; the permissible strength of causal conclusions that can be attributed to the results of an analysis often hinge on this distinction (Morgan & Winship, 2007).

The Example Data and Specification of the Models

Throughout the remainder of this volume, we use several numerical examples to illustrate a variety of multivariate linear model analyses. All the examples use continuously distributed interval level–dependent variables. The first and second data sets are used in Chapter 3 to introduce the estimation of the parameters in the multivariate general linear model. They will also be used as running examples to illustrate results on multivariate measures of strength of association (Chapter 4), multivariate test statistics (Chapter 4), and the multivariate general linear hypothesis testing procedure (Chapter 5). The third data set is used to illustrate MANOVA models, including a single-classification MANOVA and a 3 × 2 factorial MANOVA with two main effects and their interaction (Chapter 6). The first and second data sets are also used to illustrate the recovery of two of the four multivariate test statistics from only univariate quantities (Chapter 4) and to illustrate the details of canonical correlation analysis (CCA) that subsumes all the models dealt with in this volume (Chapter 7). The examples are drawn from several disciplines, including personnel psychology, anthropology, environmental epidemiology, and neuropsychology. To set the stage of model specification, the conceptual basis of each example data set is described below along with summary descriptive statistics. The specification of the analytic models appropriate for each of the examples will be a central part of subsequent chapters. The models to be specified will include MMR, MANOVA, and CCA.

Example 1: Personality and Success in the Job Application Process (MMR, CCA)

Caldwell and Burger (1998) conducted an observational study of 99 college students nearing the completion of their studies and who were anticipating entering the employment market. Individual differences on personality dimensions are thought to be among the many factors that are important in achieving a successful outcome to the job application and interviewing process. Three dimensions of personality drawn from the Five-Factor model of personality (Costa & McRae, 2000)[4] are used here to illustrate the estimation of the parameters and tests of hypotheses of an MMR model with four response variables: background preparation for the interviews, social preparation for the interviews, the number of follow-up interviews achieved, and the number of offers of employment received. For three predictor variables of Neuroticism, Extraversion, and Conscientiousness, their defining characteristics (facets) provide the conceptual bases for the predictions. The personality variable of Neuroticism is characterized by anxiety, hostility, depression, self-consciousness, impulsiveness, and vulnerability. It is easy to see how these characteristics might impede both preparation for, and success in, the job-seeking process. On the other hand, Extraversion is characterized by warmth, gregariousness, assertiveness, activity, excitement seeking, and positive emotions—all of which would predict success in the interpersonal aspects of seeking employment. The personality dimension of Conscientiousness is defined by features of competence, order, dutifulness, achievement striving, self-discipline, and deliberation—facets that may well predict variation in the careful preparation for the job interview process that should also be related to success. It can be hypothesized that a significant proportion of the joint variation in the successful outcome variables would be predictable from these personality variables. Caldwell and Burger (1998) give further details of the underlying rationale. The means, standard deviations, and correlations of the Personality–Job Application data are presented in Table 2.1.

---

[4]Caldwell and Burger (1998) present means, standard deviations, and correlations for all five of the Five-Factor personality dimensions. Neuroticism, Extraversion, and Conscientiousness were selected for predictor variables due to their theoretical relevance to the dependent variables. For this example, we generated a set of $n = 99$ fictitious data cases based on the descriptive statistics of Caldwell and Burger (1998, p. 128, Table 2), which exactly reproduced the mean, variance, and correlational structure reported in their manuscript. These fictitious data were used for the illustrative analyses presented here. The individual cases, per se, are not absolutely necessary for the analyses presented in this text. The multivariate analyses reported in this volume can be computed from intermediate statistics (e.g., means, variances, and correlations) by matrix language programs (e.g., SAS IML, SPSS MATRIX, STATA MATRIX) or by some available software packages. Harris (2001, pp. 305–307) gives instructions for multivariate analysis based on means, standard deviations, and correlations using the SPSS MANOVA procedure.

Table 2.1 Means, Standard Deviations, and Correlations for the Personality–Job Application Data

	1	*2*	*3*	*4*	*5*	*6*	*7*
1. Neuroticism	1.000						
2. Extraversion	–.100	1.000					
3. Conscientiousness	–.200	.330	1.000				
4. Background preparation	–.140	–.040	.270	1.000			
5. Social preparation	–.090	.380	.220	.420	1.000		
6. Follow-up interview	–.050	.270	.380	.200	.350	1.000	
7. Offers	–.210	.340	.050	–.140	.240	.410	1.000
Mean	25.62	38.03	39.99	13.34	11.89	0.49	0.38
SD	7.10	6.00	5.98	4.12	4.98	0.45	0.35

Source: From Caldwell & Berger's "Personality Characteristics of Job Applicants and Success in Screening Interviews" (1998), Table 1, p. 128.

Note: $n = 99$, critical values of $r = .195$ (at $\alpha = .05$) and $.254$ (at $\alpha = .01$).

Example 2: PCB Exposure, Age, Gender, Cardiovascular Disease Risk Factors, and Cognitive Functioning (MMR, CCA)

In certain areas of the United States, there is concern over the possible adverse effects of industrially produced environmental contaminants (e.g., polychlorinated biphenyls [PCBs]) on public health—both physical and psychological (Carpenter, 2006). Exposure to PCBs has been hypothesized to adversely affect measures of two related, but conceptually distinct, sets of outcome variables: two major risk factors of cardiovascular disease (physical) and three measures of cognitive functioning (neuropsychological). Because the liver is heavily involved in the body's attempt to remove toxic substances from the bloodstream (PCBs in this example), it has been hypothesized that overactivation of the liver concomitantly leads to an overproduction of cholesterol and triglycerides, which are two known major risk factors for cardiovascular disease (Goncharov et al., 2008). There is also speculation that exposure to PCBs may also have adverse effects on cognitive functioning—such as memory and cognitive flexibility (Lin, Guo, Tsai, Yang, & Guo, 2008). The data of Example 2 consist of six response variables: cholesterol, triglycerides, immediate memory, delayed memory, and two measures of cognitive flexibility (Stroop Color and Stroop Word tests), which are hypothesized to be adversely affected by exposure to PCBs. The multivariate linear model fitted to these data also includes age and gender. It is well known that liver function, memory, and cognitive flexibility are declining functions of age; assessing the effect of age on these dependent variables can provide control of inevitable confounding—since body burden of PCBs is a function of time, age is an obvious confound for any effect of exposure (e.g., $r_{PCBs.age} = .73$). We include gender as an explanatory variable in these models insofar as gender is known to be modestly related to both physical and psychological classes of dependent variable. The descriptive statistics for these example data, based on $n = 262$ cases, is shown in Table 2.2 and will be used to illustrate both MMR analysis and the related CCA.

Example 3: Stature Differences of Indigenous North American Populations (MANOVA)

Auerbach and Ruff (2010) present data on measurements of stature, relative lower limb length, and crural index[5] of skeletal pre-European indigenous

---

[5]A crural index is the ratio of the length of the tibia to the length of the femur bone. Since the data used in this example are summary statistics, the data are in the aggregate and will show less within group variability than would data based on the original 967 observations. There are both pros and cons (Lubinski & Humphreys, 1996; Robinson, 1950) surrounding the use of aggregate data; such data are more than adequate for our purposes here.

Table 2.2 Means, Standard Deviations, and Correlations for the PCB-CVD-NPSY Data

	1	*2*	*3*	*4*	*5*	*6*	*7*	*8*	*9*
1. Age	1.000								
2. Gender	.047	1.000							
3. PCBs	.731	–.130	1.000						
4. Visual memory–I	–.387	–.043	–.364	1.000					
5. Visual memory-D	–.374	.033	–.373	.779	1.000				
6. Stroop word	–.199	.145	–.169	.202	.209	1.000			
7. Stroop color	–.260	.137	–.207	.172	.193	.733	1.000		
8. Cholesterol	.359	.001	.378	–.114	–.142	–.104	–.143	1.000	
9. Triglycerides	.327	–.102	.386	–.070	–.100	–.044	–.080	.561	1.000
Mean	37.89	0.67	0.37	9.93	8.58	91.98	70.30	2.27	2.08
SD	13.47	0.47	0.37	3.29	3.60	23.98	19.72	0.09	0.25

Source: Data has appeared in Goncharov et al. (2008). *Environmental Research, 106,* 226–239; and Haase et al. (2009). *Environmental Research, 109,* 73–85.

Note: PCB = polychlorinated biphenyls; CVD = cardiovascular disease; NPSY = neuropsychological functioning. $n = 262$, critical values of $r = .10$ (at $\alpha = .05$) and .18 (at $\alpha = .01$). Visual memory-I = immediate recall, Visual memory-D = delayed recall. PCBs are log transformed.

populations of North America. Stature information is important in the study of the origin and distribution of pre-European indigenous populations in North America. From 75 different sites in North America, the authors evaluated the three variables on the skeletal remains of 535 males and 432 females. The means of the three dependent variables for males and females at each of the 75 sites provide a total sample of $n = 145$ cases as the data used in this example (see Auerbach & Ruff, 2010, Tables 1 and 2). Auerbach and Ruff have clustered these archeological sites into 11 regions based on natural (geographic) and cultural designations, and further clustered the sites into four geographically distinct groupings: (1) High Latitude Arctic Group, (2) Temperate: West Group, (3) Great Plains Group, and (4) Temperate: East Group. This clustering leads naturally to the specification of a four-group, one-way MANOVA model with three dependent variables.[6] There are at least two ways to describe the research question, formulate hypotheses, and specify the model for MANOVA designs. One common approach is to ask whether the vectors of the three dependent variable means differ simultaneously across the four clusters of sites. Specifying this hypothesis in vector notation with columns defined by $g = 4$ groups, and rows defined $p = 3$ response variables, the null hypothesis of equality of the group mean vectors is written as

$$H_0 : \boldsymbol{\mu}_1 = \boldsymbol{\mu}_2 = \boldsymbol{\mu}_3 = \boldsymbol{\mu}_4$$

or in expanded form as

$$\mathrm{H}_0 : \begin{bmatrix} \mu_{11} \\ \mu_{21} \\ \mu_{31} \end{bmatrix} = \begin{bmatrix} \mu_{12} \\ \mu_{22} \\ \mu_{32} \end{bmatrix} = \begin{bmatrix} \mu_{13} \\ \mu_{23} \\ \mu_{33} \end{bmatrix} = \begin{bmatrix} \mu_{14} \\ \mu_{24} \\ \mu_{24} \end{bmatrix}.$$

An alternate way of characterizing the one-way MANOVA is to ask if there is a significant amount of joint variation in the three dependent variables that can be accounted for by group membership. The linear model of Equation 2.1 can be specified to address this question by adopting one of several methods for coding the design matrix **X** to identify levels of a MANOVA factor contained in a categorical (qualitative) variable of group membership. The coding method can be chosen such that the parameter estimates identify differences between the means as reflected in the hypothesis above. This method of solving MANOVA problems is instructive in that the

---

[6]Auerbach and Ruff combine the temperate groups into a single cluster in their manuscript. We preserve the four-group clustering of regions for this one-way MANOVA example for pedagogical reasons.

output from the linear model most frequently associated with regression analysis (e.g., R^2) is integrated with the information most frequently associated with the classical solution to the analysis of variance (i.e., mean differences). We will undertake a more careful discussion of these equivalences in Chapter 6 and illustrate different methods of coding the design matrix to capture group differences. The means and standard deviations of the three stature response variables classified by the four site clusters of the Auerbach and Ruff data are displayed in Table 2.3. The correlations among the response variables and the grand means are given in Table 2.4.

Table 2.3 Means and (Standard Deviations) for the Four Group, One-Way MANOVA on Stature

	Mean Stature	*Mean Lower Limb Length*	*Mean Crural Index*
Group 1	153.07	48.32	81.61
High Latitude Arctic	(4.90)	(0.73)	(1.37)
Group 2	157.20	48.64	84.87
Temperate: West	(7.43)	(0.82)	(1.28)
Group 3	161.20	49.21	85.64
Great Plains Group	(6.91)	(0.60)	(1.27)
Group 4	161.60	49.12	84.59
Temperate: East	(5.84)	(0.71)	(0.96)

Source: From Auerbach & Ruff (2010). "Stature Estimation Formulae for Indigenous North American Populations", Table 1, pp. 193–194.

Note: $n_1 = 26$, $n_2 = 54$, $n_3 = 14$, $n_4 = 51$.

Table 2.4 Correlations Among the Three Response Variables for the Stature Estimation Data

	Mean Stature	*Mean Lower Limb Length*	*Mean Crural Index*
Mean stature	1.000		
Mean lower limb length	.570	1.000	
Mean crural index	.354	.265	1.000
Mean	158.39	48.81	84.27
SD	7.12	0.80	1.74

Source: From Auerbach & Ruff (2010). "Stature Estimation Formulae for Indigenous North American Populations", Table 2, pp. 195–197.

Note: Means, standard deviations, and correlations are based on the full sample of $n = 145$.

Table 2.5 Means and Standard Deviations for the 2 × 3 Factorial MANOVA

		Factor B								
		b_1			b_2			b_3		
		Y_1	Y_2	Y_3	Y_1	Y_2	Y_3	Y_1	Y_2	Y_3
Factor A	a_1	156.74	45.54	81.71	164.03	49.11	.01	167.65	49.62	85.833
		(3.41)	(0.60)	(1.13)	(4.55)	(0.66)	(1.10)	(2.10)	(0.43)	(1.60)
	a_2	149.39	48.11	81.51	154.17	48.62	84.44	154.74	48.79	85.45
		(3.02)	(0.80)	(1.61)	(5.48)	(0.86)	(1.12)	(1.34)	(0.44)	(9.92)

Note: Y_1 = stature, Y_2 = lower limb length, Y_3 = crural index. Factor A = Sex, Factor B = Geographic Cluster. Standard deviations are in parentheses.

Example 4: A 2 × 3 Factorial MANOVA—Sex by Geographic Group of the Stature Data

In addition to the regional identification of each case in the 75 North American sites, Auerbach and Ruff (2010) also catalogued their data as male or female according to the sex of the skeletal remains. Thus, the 70 sites with complete data (five sites had no females) can be partitioned into male (n = 75) and female (n = 70) groups. When the factor for sex is crossed with a factor of geographic organization—11 regions sorted into three clusters—the data can be organized into a 2 × 3 factorial analysis of variance design. In a factorial design with multiple dependent variables (i.e., stature, lower limb length, and crural index), the primary focus of the MANOVA is on the three sources of influence in the model—the main effects of sex and geographic region and the interaction between the two. While differences in the dependent variables across geographic groups (Factor A) as well as mean differences between genders (Factor B) can be important, the interpretation of the analysis may depend on the A × B interaction. Assessing if vectors of mean differences between levels of one factor are constant across the levels of the second factor is usually a major goal of factorial MANOVA. This 2 × 3 classification of the Auerbach and Ruff data provide the basis of the factorial MANOVA illustration presented in Chapter 6. As was the case with the one-way MANOVA, the design can be characterized in the classical way as tests of differences between mean vectors or as a linear model with predictor vectors designed to contrast vectors of group mean differences. The means and standard deviations of the six cells of this 2 (Sex) × 3 (Geographic Cluster) MANOVA design are displayed in Tables 2.5 and 2.6.

Table 2.6 Cell Sample Sizes for the 2 × 3 Factorial MANOVA

	b_1	b_2	b_3
a_1	$n_{11} = 13$	$n_{11} = 55$	$n_{11} = 7$
a_2	$n_{11} = 13$	$n_{11} = 50$	$n_{11} = 7$

CHAPTER 3. ESTIMATING THE PARAMETERS OF THE MULTIVARIATE GENERAL LINEAR MODEL

As introduced in Chapter 1, the univariate linear model emphasizes the prediction of a single criterion variable from one or more predictor variables. Multivariate linear models are characterized by models with more than one response variable to be analyzed as a function of one or more predictor variables. The population multivariate linear model is written as

$$\mathbf{Y}_{(n\times p)} = \mathbf{X}_{(n\times q+1)}\mathbf{B}_{(q+1\times p)} + \mathbf{E}_{(n\times p)}, \qquad [3.1]$$

in which the columns of the observed data matrix $\mathbf{Y}$ correspond to the p dependent variables. If $p = 1$, the model reduces to the univariate model of Equation 1.3. The columns of the $(n \times q + 1)$ design matrix $\mathbf{X}$ are the observed values of the predictor variables along with the $X_0 \equiv 1$ unit vector. The characteristics of the design matrix are not affected by the univariate-multivariate distinction. The $(n \times p)$ matrix of errors of the multivariate model, $\mathbf{E} = \mathbf{Y} - \mathbf{XB}$ (one column associated with each dependent variable), is obtained by subtraction and these errors can be evaluated after the $(q + 1 \times p)$ matrix of unknown population regression coefficients in $\mathbf{B}$ have been estimated. As before, n denotes the sample size of the model.

When the model contains a single criterion variable, the least squares sample estimates of the parameters of the model of Equation 1.3 were found by evaluating $\hat{\boldsymbol{\beta}} = (\mathbf{X}'\mathbf{X})^{-1}(\mathbf{X}'\mathbf{y})$. In that univariate model, the estimated regression coefficients of the optimally weighted linear combination of the predictor variables was a vector of coefficients, $\hat{\boldsymbol{\beta}}' = (\hat{\beta}_0, \hat{\beta}_1, \cdots, \hat{\beta}_q)$. Estimating the parameters of Equation 3.1 in the multivariate case by the criterion of least squares is a straightforward generalization of the univariate process that results in a $(q + 1 \times p)$ matrix of parameter estimates, $\hat{\mathbf{B}}$, which is augmented to accommodate the multiple response variables in $\mathbf{Y}$.

The parameters of the multivariate model are estimated by the criterion of least squares by selecting estimates that minimize the sum of squared errors of Equation 3.1, $\mathbf{E}'\mathbf{E}$. The $(p \times p)$ matrix of the sums and squares and cross products of the errors ($SSCP_{ERROR}$) of the multivariate model is

$$\mathbf{E}'\mathbf{E}_{(p\times p)} = (\mathbf{Y} - \mathbf{XB})'(\mathbf{Y} - \mathbf{XB}). \qquad [3.2]$$

Performing the multiplication and noting that $\mathbf{B}'\mathbf{X}'\mathbf{Y} = \mathbf{B}'\mathbf{X}'\mathbf{XB}$, the resulting expression becomes

$$\mathbf{E}'\mathbf{E}_{(p\times p)} = \mathbf{Y}'\mathbf{Y} - \mathbf{B}'\mathbf{X}'\mathbf{Y}. \qquad [3.3]$$

The criterion of least squares is employed to select the estimates of the unknown values of **B** that simultaneously minimize the error sums of squares of the p dependent variables contained on the main diagonal of Equation 3.3. The trace of $\mathbf{E'E}$ [$Tr(\mathbf{E'E})$]—the sum of the sums of squared errors—is the multivariate quantity to be minimized. Taking the matrix of partial derivatives of $\mathbf{E'E}$, setting the result equal to **0**, substituting the least squares estimator $\hat{\mathbf{B}}$ for **B**, and solving the normal equations reveal that the least squares estimates of the parameters of the multivariate model are found by[1]

$$\hat{\mathbf{B}} = (\mathbf{X'X})^{-1}\mathbf{X'Y}. \qquad [3.4]$$

The matrix of estimated regression coefficients for a multivariate model contains columns corresponding to each of the p response variables and rows for each of the $q + 1$ predictor variables and will have the form

$$\hat{\mathbf{B}}_{(q+1\times p)} = \begin{bmatrix} \hat{\beta}_{01} & \hat{\beta}_{02} & \cdots & \hat{\beta}_{0p} \\ \hat{\beta}_{11} & \hat{\beta}_{12} & \cdots & \hat{\beta}_{1p} \\ \vdots & \vdots & \ddots & \vdots \\ \hat{\beta}_{q1} & \hat{\beta}_{q2} & \cdots & \hat{\beta}_{qp} \end{bmatrix}.$$

The column vectors of $\hat{\mathbf{B}}$ are a collection of univariate models; separately estimating the p univariate regression models and collecting the p vectors into a single matrix would achieve the same result.

Substituting $\hat{\mathbf{B}}$ for **B** and $\hat{\mathbf{E}}$ for **E** in Equation 3.1, we can write out the constituent matrices of the sample multivariate linear model as

$$\begin{bmatrix} Y_{11} & Y_{12} & \cdots & X_{1p} \\ Y_{21} & Y_{22} & \cdots & Y_{2p} \\ \vdots & \vdots & \ddots & \vdots \\ Y_{n1} & Y_{n2} & \cdots & Y_{np} \end{bmatrix} = \begin{bmatrix} 1 & X_{11} & X_{12} & \cdots & X_{1q} \\ 1 & X_{21} & X_{22} & \cdots & X_{2q} \\ \vdots & \vdots & \vdots & \ddots & \vdots \\ 1 & X_{n1} & X_{n2} & \cdots & X_{nq} \end{bmatrix}$$

$$\begin{bmatrix} \hat{\beta}_{01} & \hat{\beta}_{02} & \cdots & \hat{\beta}_{0p} \\ \hat{\beta}_{11} & \hat{\beta}_{12} & \cdots & \hat{\beta}_{1p} \\ \hat{\beta}_{21} & \hat{\beta}_{22} & \cdots & \hat{\beta}_{2p} \\ \vdots & \vdots & \ddots & \vdots \\ \hat{\beta}_{q1} & \hat{\beta}_{n2} & \cdots & \hat{\beta}_{qp} \end{bmatrix} + \begin{bmatrix} \hat{\varepsilon}_{11} & \hat{\varepsilon}_{12} & \cdots & \hat{\varepsilon}_{n1} \\ \hat{\varepsilon}_{21} & \hat{\varepsilon}_{22} & \cdots & \hat{\varepsilon}_{n2} \\ \vdots & \vdots & \ddots & \vdots \\ \hat{\varepsilon}_{n1} & \hat{\varepsilon}_{n1} & \cdots & \hat{\varepsilon}_{np} \end{bmatrix}$$

[1]Details of the derivation of Equation 3.4 are given in Rencher (1998, pp. 280–282).

with fitted values of $\widehat{\mathbf{Y}}=\mathbf{X}\widehat{\mathbf{B}}$ and estimated residuals of the fitted model $\widehat{\mathbf{E}}=\mathbf{Y}-\widehat{\mathbf{Y}}=\mathbf{Y}-\mathbf{X}\widehat{\mathbf{B}}$.

The response variables in **Y** and the estimated errors of these multivariate models in $\widehat{\mathbf{E}}$ are assumed to be random variables with underlying multivariate probability distributions, while the values of **X** and the coefficients in **B** are assumed to be fixed constants. No assumptions are *required* to compute the least squares estimates $\widehat{\mathbf{B}}$, but if certain assumptions are met the estimates are said to be the "best linear unbiased estimates (BLUE)" of the population parameters, and certain assumptions are needed to justify the validity of the multivariate test statistics developed in Chapter 5. These assumptions help ensure that the estimates are unbiased (i.e., $E(\widehat{\mathbf{B}})=\mathbf{B}$) and have the smallest sampling variance of any such linear estimates that could be obtained. There are four such assumptions: linearity, constant variance, independence, and multivariate normality, which parallel the assumptions required for the univariate model.

Linearity. The model $\mathbf{Y}=\mathbf{XB}+\mathbf{E}$ is assumed to be linear in the parameters with expected value $E(\mathbf{Y})=\mathbf{XB}$; concomitantly the errors of the model are assumed to have zero means, $E(\mathbf{E})=\mathbf{0}$. These features of the model assume that the model is correct and that no important predictor variables have been omitted from the model. If there are additional important predictors not included in the model, then their variances are absorbed in the errors whose expected value is no longer **0**.

Constant Variance. In the linear model, the variances and covariances of the errors are assumed to be constant across all cases or participants. This assumption can be described in two parts. Consider any row vector of the multivariate model that is due to a single case, $\mathbf{y}_i=\mathbf{x}_i\mathbf{B}+\boldsymbol{\epsilon}_i$. The constant variance requirement assumes that the *p* variances of $\mathbf{y}_i$ (or of $\boldsymbol{\epsilon}_i$) for a single participant are constant across all the response variables. Second, the *p* variances for each of the *n* cases are assumed to be equal and hence have the same constant variance structure.

Independence. The responses (and the errors) for each case in the model are assumed to be independent of one another. This constant covariance requirement assumes that the covariances of the *p* response variables between the *i*th and the *j*th case (for all $i \neq j$) are zero—that is, the responses of any case are assumed to be uncorrelated with any other case and the covariances, like the variances, are assumed to be constant across all cases.

The assumptions of constant variance and independence are often collected into a single $p \times p$ matrix, $\boldsymbol{\Sigma}$, of variances and covariances (0) that are constant across all participants. If the assumptions hold in the population, then $\mathbf{\Sigma}$ has the general form,

$$\mathbf{\Sigma}_{(p \times p)} = \begin{bmatrix} \sigma_1^2 & \sigma_{12} & \cdots & \sigma_{1p} \\ \sigma_{21} & \sigma_2^2 & \cdots & \sigma_{2p} \\ \vdots & \vdots & \ddots & \vdots \\ \sigma_{p1} & \sigma_{p2} & \cdots & \sigma_p^2 \end{bmatrix} = \begin{bmatrix} \sigma_1^2 & 0 & \cdots & 0 \\ 0 & \sigma_2^2 & \cdots & 0 \\ \vdots & \vdots & \ddots & \vdots \\ 0 & 0 & 0 & \sigma_p^2 \end{bmatrix}.$$

Multivariate Normality. The assumption of multivariate normality justifies the validity of the multivariate test statistics to be used for testing hypotheses on the elements of **B**. Test statistics in multivariate analysis must depend on some underlying probability distribution. In the multivariate domain, the multivariate normal distribution is the probability distribution most frequently adopted to provide a basis for testing hypotheses. For the multivariate test statistics to be valid, it is assumed that each row $\mathbf{y}_i$ of the data matrix **Y** is assumed to be distributed as multivariate normal with mean **XB** and variance-covariance matrix $\mathbf{\Sigma}$,

$$\mathbf{y}_i \sim N_p(\mathbf{XB}, \mathbf{\Sigma}). \qquad [3.5]$$

Since the distribution of $\mathbf{y}_i$ applies equally to ϵ_i the multivariate normal distribution with mean **0** and variance-covariance matrix $\mathbf{\Sigma}$ applies equally to the model errors,

$$\epsilon_i \sim N_p(\mathbf{0}, \mathbf{\Sigma}). \qquad [3.6]$$

An unbiased estimator of $\mathbf{\Sigma}$ is obtained by substituting the parameter estimates $\hat{\mathbf{B}}$ into Equation 3.3 and converting the $SSCP_{ERROR}$ matrix to a variance-covariance matrix,

$$\hat{\mathbf{\Sigma}} = \frac{1}{n-q-1}\mathbf{E}'\mathbf{E}. \qquad [3.7]$$

Further discussion of the multivariate normal distribution and its importance in multivariate analysis can be found in Tatsuoka (1988, Chap. 4), and methods for testing the assumptions of the multivariate model are given in Stevens (2007, Chap. 6). Estimating the parameters of linear models with $p > 1$ response variables can best be illustrated by numerical examples.

Example 1: Personality Characteristics and Job Application Success

We continue here with the personality data with $p = 4$ response variables that measure the degree of success in application for employment: background preparation, social preparation, invitations to a second interview, and job offers. These markers of success in the job application process are modeled as a function of the intercept (unit vector $X_0 \equiv 1$) and three explanatory personality variables: neuroticism, extraversion, and conscientiousness as defined in Chapter 2. These dimensions of personality are considered by psychologists to represent enduring characteristics of the individual that help govern behavior and may have implications for how these three factors might usefully predict preparation for, behavior during, and success in the job application and interview process.

The data are multivariate with p = 4 response variables, $\mathbf{Y}_{(n \times p)} = [\mathbf{y}_1, \mathbf{y}_2, \mathbf{y}_3, \mathbf{y}_4]$ collected over the course of the experimental period. To account for the joint variability in this multivariate set of dependent variables as a function of the variability in the $q = 3$ predictor variables requires an estimate of the parameters of the model. We estimate the parameters of the model specified by solving for $\widehat{\mathbf{B}}$ in Equation 3.4. With $n = 99$ observations, the data matrix $\mathbf{Y}_{(n \times p)}$ is of order (99 × 4) consisting of the four continuously distributed response variables. The design matrix $\mathbf{X}_{(n \times q+1)}$ is of order (99 × 4) with four columns consisting of $X_0 \equiv 1$ to estimate the intercept along with the three explanatory personality measures. The summary descriptive statistics and the matrix of correlations among the variables for this example were presented in Table 2.1. The general form of the data matrices, **Y** and **X**, shown here are the usual beginning point of the analysis,

$$\mathbf{Y}_{(99 \times 4)} = \begin{bmatrix} Y_{11} & Y_{12} & Y_{13} & Y_{14} \\ Y_{21} & Y_{22} & Y_{23} & Y_{24} \\ Y_{31} & Y_{32} & Y_{33} & Y_{34} \\ \vdots & \vdots & \vdots & \vdots \\ Y_{99\,1} & Y_{99\,2} & Y_{99\,3} & Y_{99\,4} \end{bmatrix} = \begin{bmatrix} 18 & 13 & 1 & 1 \\ 13 & 12 & 1 & 1 \\ 11 & 5 & 2 & 0 \\ \vdots & \vdots & \vdots & \vdots \\ 12 & 14 & 2 & 1 \end{bmatrix},$$

$$\mathbf{X}_{(99 \times 4)} = \begin{bmatrix} 1 & X_{11} & X_{12} & X_{13} \\ 1 & X_{21} & X_{22} & X_{23} \\ 1 & X_{31} & X_{11} & X_{11} \\ \vdots & \vdots & \vdots & \vdots \\ 1 & X_{99\,1} & X_{99\,2} & X_{99\,3} \end{bmatrix} = \begin{bmatrix} 1 & 25 & 34 & 44 \\ 1 & 29 & 34 & 40 \\ 1 & 19 & 35 & 35 \\ \vdots & \vdots & \vdots & \vdots \\ 1 & 25 & 42 & 42 \end{bmatrix}.$$

A graphic view of the relationships between the three predictors and four criterion variables is shown in Figure 3.1, which correspond to the zero-order correlations of Table 2.1.

Following Equation 3.4, the estimation of the parameter matrix $\mathbf{B}_{(q+1 \times p)}$, which is of order (4 × 4) for this example, is obtained by first evaluating the matrix of the *uncorrected* sums of squares and cross-products (SSCP) of the predictor variables, $\mathbf{X'X}_{(4\times 4)}$, and the *uncorrected* sums of cross products (SCP) between predictors and response variables, $\mathbf{X'Y}_{(4\times 4)}$.[2] Computing these matrices, we have

$$\mathbf{X'X} = \begin{bmatrix} 1 & 1 & 1 & \cdots & 1 \\ X_{11} & X_{21} & X_{31} & \cdots & X_{99\,1} \\ X_{12} & X_{22} & X_{32} & \cdots & X_{99\,2} \\ X_{13} & X_{23} & X_{33} & \cdots & X_{99\,3} \end{bmatrix} \begin{bmatrix} 1 & X_{11} & X_{12} & X_{13} \\ 1 & X_{21} & X_{22} & X_{23} \\ 1 & X_{31} & X_{11} & X_{11} \\ \vdots & \vdots & \vdots & \vdots \\ 1 & X_{99\,1} & X_{99\,2} & X_{99\,3} \end{bmatrix}$$

$$= \begin{bmatrix} 99.00 & 2536.38 & 3764.97 & 3959.01 \\ 2536.38 & 69922.24 & 96041.05 & 151721.51 \\ 3764.97 & 96041.05 & 14679.81 & 151721.51 \\ 3959.01 & 100597.51 & 151721.51 & 161825.33 \end{bmatrix}$$

Figure 3.1 Relationships of Three Personality Measures and Four Job Application Outcomes

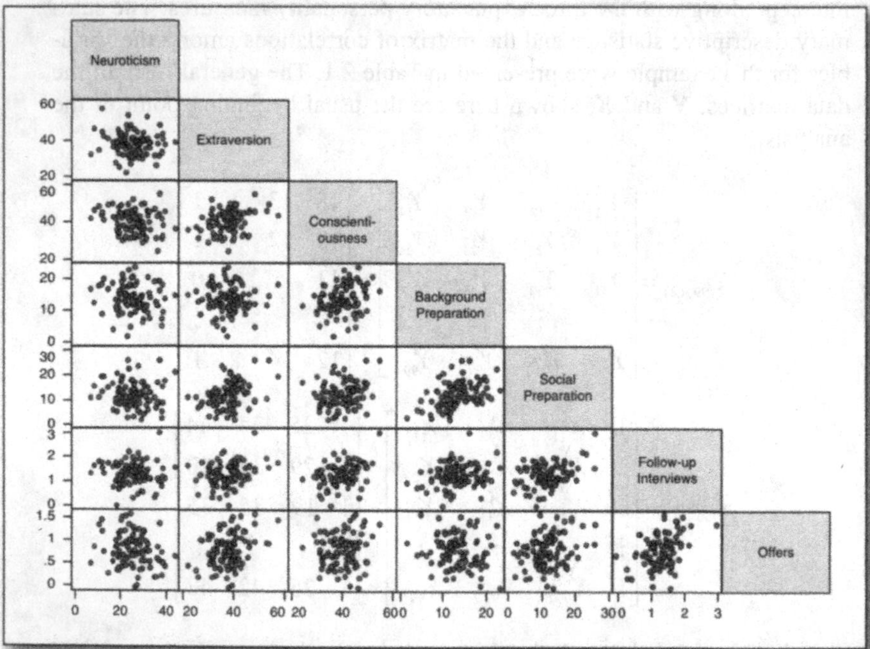

[2]The uncorrected sums of squares and cross products are based on summing and squaring the raw data for the variables in **Y** and **X**. In later sections, it will be more convenient to deal with mean corrected sums of squares and cross-product (SSCP) matrices that are better suited for many statistical applications.

and

$$\mathbf{X'Y} = \begin{bmatrix} 1 & 1 & 1 & \cdots & 1 \\ X_{11} & X_{21} & X_{31} & \cdots & X_{99\,1} \\ X_{12} & X_{22} & X_{32} & \cdots & X_{99\,2} \\ X_{13} & X_{23} & X_{33} & \cdots & X_{99\,3} \end{bmatrix} \begin{bmatrix} Y_{11} & Y_{12} & Y_{13} & Y_{14} \\ Y_{21} & Y_{22} & Y_{23} & Y_{24} \\ Y_{31} & Y_{32} & Y_{33} & Y_{34} \\ \vdots & \vdots & \vdots & \vdots \\ Y_{99\,1} & Y_{99\,2} & Y_{99\,3} & Y_{99\,4} \end{bmatrix}$$

$$= \begin{bmatrix} 1320.66 & 1177.11 & 48.51 & 37.62 \\ 33433.97 & 29845.70 & 1227.17 & 912.68 \\ 50127.80 & 45878.22 & 1916.28 & 1500.66 \\ 53465.10 & 47714.69 & 2040.13 & 1514.68 \end{bmatrix}.$$

Postmultiplying $\mathbf{X'Y}$ by the inverse of $\mathbf{X'X}$ yields the estimates of the parameters

$$\hat{\mathbf{B}} = (\mathbf{X'X})^{-1}\mathbf{X'Y} = \begin{bmatrix} 0.93912 & -0.00765 & -0.0082 & -0.01050 \\ -0.00765 & 0.00021 & 0.00001 & 0.00005 \\ -0.00823 & 0.00001 & 0.00032 & -0.00010 \\ -0.01050 & 0.00005 & -0.00010 & 0.00033 \end{bmatrix} \times \begin{bmatrix} 1320.66 & 1177.11 & 48.51 & 37.62 \\ 33433.97 & 29845.70 & 1227.17 & 912.68 \\ 50127.80 & 45878.22 & 1916.28 & 1500.66 \\ 53465.10 & 47714.69 & 2040.13 & 1514.68 \end{bmatrix}$$

with the result for the personality data being,

$$\hat{\mathbf{B}}_{(4\times4)} = \begin{array}{c} \\ X_0 \\ X_1 \\ X_2 \\ X_3 \end{array} \begin{array}{c} \begin{matrix} Y_1 & Y_2 & Y_3 & Y_4 \end{matrix} \\ \begin{bmatrix} \hat{\beta}_{01} & \hat{\beta}_{02} & \hat{\beta}_{03} & \hat{\beta}_{04} \\ \hat{\beta}_{11} & \hat{\beta}_{12} & \hat{\beta}_{13} & \hat{\beta}_{14} \\ \hat{\beta}_{21} & \hat{\beta}_{22} & \hat{\beta}_{23} & \hat{\beta}_{24} \\ \hat{\beta}_{31} & \hat{\beta}_{32} & \hat{\beta}_{33} & \hat{\beta}_{34} \end{bmatrix} \end{array} = \begin{bmatrix} 10.361 & -1.626 & -1.031 & 0.088 \\ -0.055 & -0.025 & 0.002 & -0.010 \\ -0.102 & 0.285 & 0.012 & 0.021 \\ 0.207 & 0.083 & 0.025 & -0.006 \end{bmatrix}.$$

The four columns of $\hat{\mathbf{B}}$ are the estimates of the regression coefficients for X_0, X_1, X_2, and X_3 that optimally predict the values of Y_1, Y_2, Y_3, and Y_4 from the fitted multivariate regression model. The matrix solution is compact and efficient, but produces nothing more than the four sets of regression coefficients that would have been found by regressing Y_1, Y_2, Y_3, and Y_4 separately on the unit vector and the three predictor variables. By extension the interpretation of the coefficients in the multivariate model are identical to the descriptions that would be applied separately to the univariate coefficients for each response. Each of the columns of $\hat{\mathbf{B}}$ represents a full univariate regression solution to one of the dependent variables. For the example, the model for the predicted values of background preparation is

$$\hat{Y}_{bkg.prep} = 10.361 - .055(Neuroticism) - .102(Extraversion) \\ + .207(Conscientiousness),$$

in which Neuroticism and Extraversion are negatively related to background preparation and Conscientiousness is positively related and the largest of the three coefficients. Conversely, for the number of follow-up interviews,

$$\hat{Y}_{follow\text{-}up} = -1.031 + .002(Neuroticism) + .012(Extraversion) \\ + .025(Conscientiousness).$$

Neuroticism, Extraversion, and Conscientiousness all positively influence the number of follow-up interviews received, the largest being the coefficient for Conscientiousness. Similar interpretations apply to the vectors of coefficients for social preparation and offers received:

$$\hat{Y}_{social.prep} = -1.626 + .025(Neuroticism) + .285(Extraversion) \\ + .083(Conscientiousness)$$

for which Extraversion is the largest unique contributor to that outcome, and to offers received,

$$\hat{Y}_{offers} = .088 - .010(Neuroticism) + .021(Extraversion) \\ - .006(Conscientiousness)$$

for which Extraversion is largest of the three predictors.

The usual interpretations of these unstandardized regression coefficients apply to each of these vectors. For example, a one-unit increase in the jth predictor is accompanied by a $\hat{\beta}_{kj}$ unit increase in the kth criterion variable. Multivariate linear model coefficients, like their partial univariate counterparts, are adjusted for the relationship between any predictor variable and the remaining predictors in the model.

While one can obtain a rough idea as to the relative contribution of each explanatory variable by examining these unstandardized coefficients within a specific response variable,[3] it is difficult to make comparative statements

[3]The standard deviations of Neuroticism, Extraversion, and Conscientiousness (Table 2.1) are roughly equal and comparative statements about regression coefficients within any response variable are justified. The opposite is true for the response variables of this example that are on very different scales of measurement.

about predictors across response variables because the scales of measurement of the variables in $\mathbf{Y}$ are not comparable, as documented by the differences in the standard deviations of Table 2.1. Estimating the model in standard score form may alleviate this difficulty of interpretation in the sense that the parameters will be estimated on a common metric.[4]

Estimating the Parameters of the Multivariate Linear Model in Standard Score Form

As in the univariate case, different metrics can be employed when analyzing multivariate data. Estimating the parameters of the model in standard score form provides a second interpretive framework that circumvents the problems of interpreting unstandardized coefficients that differ in scale for both the Y and X variables. An equivalent model to that of Equation 3.1, but expressed in standard score form, can be solved in terms of the relevant correlation matrices between the Y and X variables that specify the multivariate problem

$$\mathbf{Z}_Y = \mathbf{Z}_X\mathbf{B}^* + \mathbf{E}^*, \qquad [3.8]$$

where $\mathbf{Z}_Y$ and $\mathbf{Z}_X$ are the matrices of standardized response and predictor variables transformed to mean zero and unit variance, $\mathbf{B}^*$ is a matrix of standardized regression coefficients and $\mathbf{E}^*$ is a matrix of standardized errors of prediction. The error sums of squares and cross-products of Equation 3.8 are found as

$$\mathbf{E}^{*\prime}\mathbf{E}^* = \left(\mathbf{Z}_Y - \mathbf{Z}_X\mathbf{B}^*\right)'\left(\mathbf{Z}_Y - \mathbf{Z}_X\mathbf{B}^*\right) = \mathbf{Z}_Y'\mathbf{Z}_Y - \mathbf{B}^{*\prime}\mathbf{Z}_X'\mathbf{Z}_Y. \qquad [3.9]$$

Substituting the sample estimates $\hat{\mathbf{B}}^*$ for the population parameters $\mathbf{B}^*$ of Equation 3.9, taking the partial derivatives of the trace of $\mathbf{E}^{*\prime}\mathbf{E}^*$ (see Timm, 1975, pp. 308–311), setting the result equal to zero to minimize the $Tr(\mathbf{E}^{*\prime}\mathbf{E}^*)$, and solving the normal equations yield the least squares parameter estimates in standard score form,

$$\hat{\mathbf{B}}^* = (\mathbf{Z}_X'\mathbf{Z}_X)^{-1}\mathbf{Z}_X'\mathbf{Z}_Y. \qquad [3.10]$$

The estimates of the parameters of Equation 3.10 can also be obtained by reference only to the correlation matrices that underlie the standard score

---

[4]The alternative interpretations of unstandardized and standardized regression coefficients discussed in earlier chapters apply equally to multivariate multiple regression problems.

solution. Dividing a standardized SSCP matrix by $n - 1$ defines the sample correlation matrix among its variables. Hence,

$$\mathbf{R}_{YY} = \frac{1}{n-1}\mathbf{Z}'_Y\mathbf{Z}_Y,\ \mathbf{R}_{XX} = \frac{1}{n-1}\mathbf{Z}'_X\mathbf{Z}_X,\ \mathbf{R}_{YX} = \frac{1}{n-1}\mathbf{Z}'_Y\mathbf{Z}_X,$$

and

$$\mathbf{R}_{XY} = \mathbf{R}'_{YX} = \frac{1}{n-1}\mathbf{Z}'_X\mathbf{Z}_Y.$$

It can be shown that the estimates of the model parameters in standard score form can be conveniently obtained by

$$\widehat{\mathbf{B}}^* = \mathbf{R}_{XX}^{-1}\mathbf{R}_{XY}, \tag{3.11}$$

where $\mathbf{R}_{XX}^{-1}$ is the inverse of the correlation matrix among the X variables.

The four correlation matrices $\mathbf{R}_{YY}, \mathbf{R}_{YX}, \mathbf{R}_{XX}$, and $\mathbf{R}_{XY}$ for the personality example data are

$$\mathbf{R}_{YY} = \begin{bmatrix} 1.00 & .41 & .24 & -.14 \\ .41 & 1.00 & .35 & .20 \\ .24 & .35 & 1.00 & .42 \\ -.14 & .20 & .42 & 1.00 \end{bmatrix},\ \mathbf{R}_{YX} = \begin{bmatrix} -.21 & .34 & .05 \\ -.05 & .27 & .38 \\ -.09 & .38 & .22 \\ -.14 & -.04 & .27 \end{bmatrix},$$

$$\mathbf{R}_{XX} = \begin{bmatrix} 1.00 & -.10 & -.20 \\ -.10 & 1.00 & .33 \\ -.20 & .33 & 1.00 \end{bmatrix}, \mathbf{R}_{XY} = \begin{bmatrix} -.21 & -.05 & -.09 & -.14 \\ .34 & .27 & .38 & -.04 \\ .05 & .38 & .22 & .27 \end{bmatrix}.$$

Applying Equation 3.11, the $q \times p$ matrix of standardized regression coefficients is found to be

$$\widehat{\mathbf{B}}^* = \mathbf{R}_{XX}^{-1}\mathbf{R}_{XY} = \begin{bmatrix} 1.00 & -.10 & -.20 \\ -.10 & 1.00 & .33 \\ -.20 & .33 & 1.00 \end{bmatrix}^{-1} \begin{bmatrix} -.21 & -.05 & -.09 & -.14 \\ .34 & .27 & .38 & -.04 \\ .05 & .38 & .22 & .27 \end{bmatrix}$$

$$= \begin{bmatrix} -.095 & -.036 & .033 & -.196 \\ -.148 & .344 & .164 & .356 \\ .300 & .099 & .333 & -.106 \end{bmatrix} = \begin{bmatrix} \hat{\beta}^*_{11} & \hat{\beta}^*_{12} & \hat{\beta}^*_{13} & \hat{\beta}^*_{14} \\ \hat{\beta}^*_{21} & \hat{\beta}^*_{22} & \hat{\beta}^*_{23} & \hat{\beta}^*_{24} \\ \hat{\beta}^*_{31} & \hat{\beta}^*_{32} & \hat{\beta}^*_{33} & \hat{\beta}^*_{34} \end{bmatrix}.$$

The regression coefficients contained in $\hat{\mathbf{B}}^*$ are the same standardized regression coefficients for Y_1, Y_2, Y_3, and Y_4 that would have been obtained by fitting four separate univariate regression models and they therefore have the same interpretations as the univariate estimates; each coefficient represents a $\hat{\beta}_j^*$ standard deviation change in Y per 1 *SD* change in X_j. While being mindful of the issues involved in interpreting standardized coefficients (Kim & Ferree, 1981), some useful insights can be gleaned from examining the rows and columns of $\hat{\mathbf{B}}^*$. The dependent variables in this $p = 4$ multivariate problem are also standardized, and comparisons of coefficients across columns (i.e., across response variables) are interpretable as are comparisons of coefficients within columns. For example, Neuroticism, which is characterized by worry, anxiety, hostility, and depressive symptoms, seems to have its largest negative effect on the number of offers ultimately received (–.196). Conversely, Extraversion, which is characterized by gregariousness, sociability, warmth, and positive emotions, is most heavily implicated in social preparation (.344) and the number of offers ultimately received (.356). Conscientiousness, whose central traits are competence, order, discipline, and duty, predicts background preparation (.300) and receipt of follow-up interviews (.333) far more than it predicts social preparation and offers received. It appears that there is a logically discernable pattern of how these personality characteristics differentially affect both preparation and success in the job interviewing process. The issue of how statistically substantial (i.e., significant) these effects are is a topic we discuss in subsequent chapters.

Example 2: The PCB-CVD Risk Factor: Cognitive Functioning Data

The polychlorinated biphenyls (PCB) data, introduced in Table 2.2, provide a second example of estimation of the parameters of the multivariate multiple regression model. To illustrate the procedures, we fit two models to these data—the model based on the raw scores, $\mathbf{Y} = \mathbf{XB} + \mathbf{E}$, and the model based on the standardized measures, $\mathbf{Z}_Y = \mathbf{Z}_X\mathbf{B}^* + \mathbf{E}^*$. For each model, we estimate the parameters by Equations 3.4 and 3.11, respectively.

The data described in Table 2.2 for this multivariate model consists of six response variables collected on 262 cases yielding the data matrix, $\mathbf{Y}_{(262 \times 6)}$, whose column vectors are the measures of memory function (immediate and delayed visual memory), cognitive flexibility (Stroop word and color), and two risk factors for cardiovascular disease (CVD; the serum lipids of cholesterol and triglycerides). The design matrix is a $262 \times 3 + 1$ matrix, $\mathbf{X}_{(262 \times 3 + 1)}$, which includes the unit vector and the observations of age, gender, and PCBs (log transformed). The data matrices would take the following form:

$$\mathbf{Y}_{(262\times 6)} = \begin{bmatrix} Y_{11} & Y_{12} & Y_{13} & Y_{14} & Y_{15} & Y_{16} \\ Y_{21} & Y_{22} & Y_{23} & Y_{24} & Y_{15} & Y_{16} \\ Y_{31} & Y_{32} & Y_{33} & Y_{34} & Y_{15} & Y_{16} \\ \vdots & \vdots & \vdots & \vdots & \vdots & \vdots \\ Y_{262\,1} & Y_{262\,2} & Y_{262\,3} & Y_{262\,4} & Y_{262\,5} & Y_{262\,6} \end{bmatrix} = \begin{bmatrix} 14 & 13 & 106 & 66 & 2.2 & 1.9 \\ 13 & 12 & 81 & 73 & 2.1 & 1.7 \\ 9 & 9 & 112 & 87 & 2.2 & 1.8 \\ \vdots & \vdots & \vdots & \vdots & \vdots & \vdots \\ 5 & 5 & 6 & 6 & 2.3 & 2.3 \end{bmatrix}$$

$$\mathbf{X}_{(262\times 4)} = \begin{bmatrix} 1 & X_{11} & X_{12} & X_{13} \\ 1 & X_{21} & X_{22} & X_{23} \\ 1 & X_{31} & X_{11} & X_{11} \\ \vdots & \vdots & \vdots & \vdots \\ 1 & X_{262\,1} & X_{262\,2} & X_{262\,3} \end{bmatrix} = \begin{bmatrix} 1 & 25 & 1 & -.51 \\ 1 & 20 & 1 & -.48 \\ 1 & 21 & 1 & -.48 \\ \vdots & \vdots & \vdots & \vdots \\ 1 & 58 & 0 & 1.40 \end{bmatrix}.$$

A visual sense of the direction and strength of the 18 bivariate relationships between explanatory and response variables can be seen in Figures 3.2 and 3.3.

Examination of the box plots of Figure 3.3 suggests that gender is not likely to be the most important predictor in these models; the mean difference for each of the dependent variables appears to be modest at best. Conversely, both PCBs and age appear to have stronger nonzero relationships with most of the response variables as reflected in the apparent nonzero slopes and the elliptical cluster of the data points around the possible regression lines displayed in Figure 3.2. The relative importance of these predictors to different response variables is of some interest in this research. The first step in assessing these relationships is the estimation of the parameters. The evaluation of the hypothesized effects of PCB exposure on the response variables can best be evaluated after adjusting for age and gender as both are known to be separately predictive of the response variables as well as significantly related to the level of PCB exposure. The bivariate graphs of Figure 3.2 do not clearly reflect the possible confounding of these predictor variables. These estimated coefficients of $\widehat{\mathbf{B}}$ are the *partial* regression coefficients of the predictors on each of the six criterion variables of the problem with each predictor adjusted for the remaining predictor variables in the model. For the PCB data, the intermediate uncorrected SSCP matrices are

$$\mathbf{Y} = \begin{bmatrix} 28651.00 & 24716.00 & 243400.00 & 185766.00 & 5905.17 & 5395.14 \\ 24716.00 & 22651.00 & 211378.00 & 161544.00 & 5097.12 & 4650.31 \\ 243400.00 & 211378.00 & 2366745.00 & 1784737.00 & 54736.00 & 50056.88 \\ 185766.00 & 161544.00 & 1784737.00 & 1396401.00 & 41813.97 & 38208.99 \\ 5905.17 & 5097.12 & 54736.00 & 41813.97 & 1356.55 & 1242.34 \\ 5395.14 & 4650.31 & 50056.88 & 38208.99 & 1242.34 & 1149.77 \end{bmatrix},$$

gure 3.2 Relationships of Polychlorinated Biphenyls (PCBs) and Age to Memory, Cognitive Flexibility, and Serum Lipids

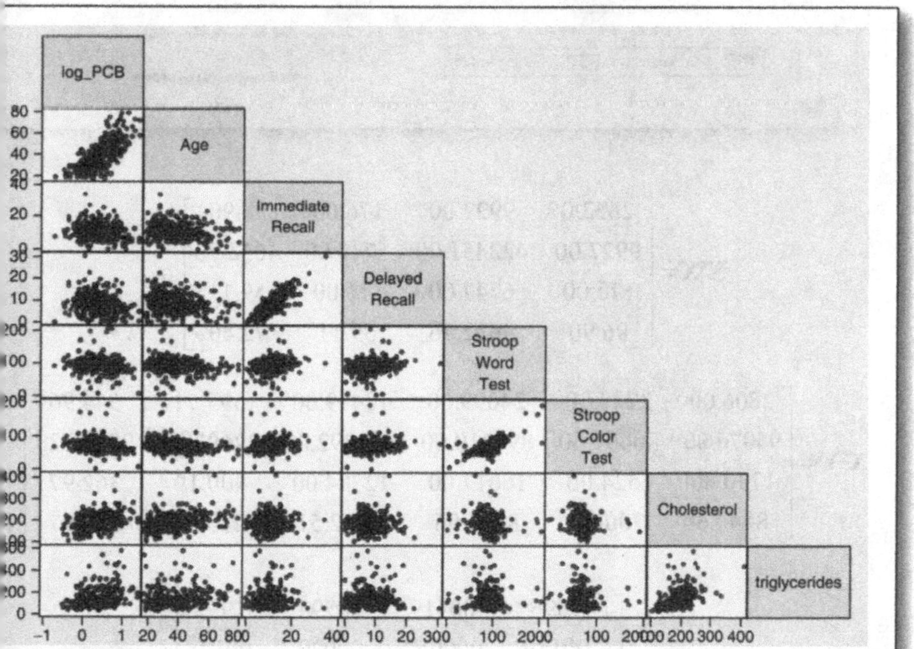

e: Polychlorinated biphenyls (PCBs), cholesterol, and triglycerides are log transformed.

Figure 3.3 Relationships of Gender to Memory, Cognitive Flexibility, and Serum Lipids

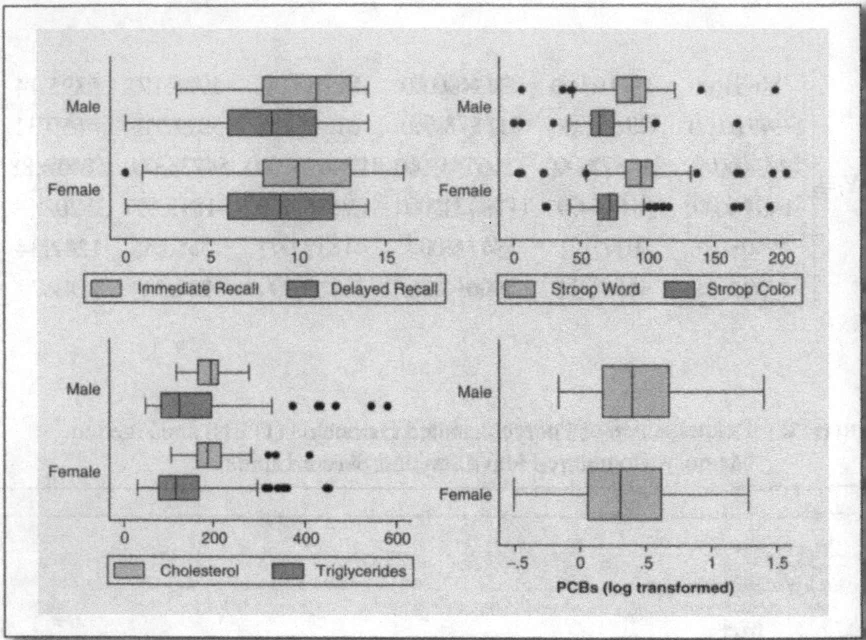

$$\mathbf{X'X} = \begin{bmatrix} 262.00 & 9927.00 & 176.00 & 96.90 \\ 9927.00 & 423451.00 & 6747.00 & 4632.76 \\ 176.00 & 6747.00 & 176.00 & 59.11 \\ 96.90 & 4632.76 & 59.11 & 72.40 \end{bmatrix},$$

$$\mathbf{X'Y} = \begin{bmatrix} 2601.00 & 2247.00 & 24099.00 & 18419.00 & 595.71 & 544.96 \\ 94076.00 & 80407.00 & 896310.00 & 679892.00 & 22683.54 & 20935.20 \\ 1730.00 & 1524.00 & 16617.00 & 12704.00 & 400.19 & 362.97 \\ 844.86 & 700.03 & 8515.96 & 6412.51 & 223.61 & 210.97 \end{bmatrix},$$

$$(\mathbf{X'X})^{-1} = \begin{bmatrix} .04696 & -.00119 & -.00794 & .01993 \\ -.00119 & .00005 & -.00020 & -.00128 \\ -.00794 & -.00020 & .01843 & .00819 \\ .01993 & -.00128 & .00819 & .06234 \end{bmatrix}.$$

From these intermediate results, the estimates of unstandardized regression coefficients of Equation 3.4 are found to be

$$\hat{\mathbf{B}}_{(4\times 6)} = \begin{bmatrix} 13.065 & 11.489 & 100.679 & 81.158 & 2.205 & 1.950 \\ -0.059 & -0.060 & -0.373 & -0.407 & 0.001 & 0.002 \\ -0.396 & 0.127 & 7.947 & 6.367 & 0.005 & -0.037 \\ -1.715 & -1.993 & 0.248 & 0.823 & 0.061 & 0.194 \end{bmatrix}.$$

The predicted values for the cases on this set of **Y** variables would be found as $\hat{\mathbf{Y}} = \mathbf{X}\hat{\mathbf{B}}$, with errors of prediction estimated by $\hat{\mathbf{E}} = \mathbf{Y} - \hat{\mathbf{Y}}$. Each column of $\hat{\mathbf{B}}_{(4\times 6)}$ contains an estimated regression equation for one of the six dependent variables. For example, the fitted model for the variables of visual memory immediate recall (Column 1 of $\hat{\mathbf{B}}$) and log-transformed cholesterol variable (Column 6 of $\hat{\mathbf{B}}$) are

$$\hat{Y}_{vm.imm} = 13.07 - 0.059(Age) - 0.396(Gender) - 1.715(\log\_PCBs)$$

$$\hat{Y}_{log\_chol} = 2.21 + 0.001(Age) + 0.005(Gender) + 0.061(\log\_PCBs).$$

The usual interpretations apply to each column of $\hat{\mathbf{B}}$. Each year of age predicts a 0.059-unit drop in immediate memory recall (about 0.4% of its range), average memory for males is predicted to be 0.396 units less than the average for females, and each $\log_{10}$ unit increase in PCBs results in a predicted drop in memory of 1.715 units (about 11% of its range). In similar fashion, $\log_{10}$ changes in cholesterol are seen to increase by 0.001 units per year of advancing age (about 0.2% of its range), average cholesterol for males is about 0.005 $\log_{10}$ units higher than for females, and a one $\log_{10}$ unit change in PCB body burden predicts a 0.061 $\log_{10}$ unit increase in cholesterol (about 10% of its range). Similar interpretations apply to the remaining four response variables.

For the same reasons discussed in previous examples, the *relative* importance of any explanatory variable across response variables, or the *relative* influence of the explanatory variables within a single response variable, is difficult to judge because elements on the same row, or the same column, of $\hat{\mathbf{B}}$ are not necessarily on the same scales and as a consequence are not directly comparable.[5] An alternative strategy for assessing the relative

[5]Transforming variables to $\log_{10}$ units to achieve greater symmetry of the distribution of sample observations also complicates the interpretation of the unstandardized coefficients.

importance of the predictors that obviates the difficulties of dissimilar scales of measurement can be approached by estimating the parameters of the model in the standard score form. The standardized coefficients of the model for the PCB data can be estimated by the correlations contained in Table 2.2. From $\mathbf{R}_{XX}$ we compute $\mathbf{R}_{XX}^{-1}$,

$$\mathbf{R}_{XX}^{-1} = \begin{bmatrix} 2.24586 & -0.32574 & -1.68366 \\ -0.32574 & 1.06448 & 0.37660 \\ -1.68366 & 0.37660 & 2.27942 \end{bmatrix},$$

Using the 3 × 6 matrix $\mathbf{R}_{XX}$ of Table 2.2,

$$\mathbf{R}_{XY} = \begin{bmatrix} .387 & -.347 & -.199 & -.259 & .359 & .327 \\ -.043 & .033 & .146 & .137 & .001 & -.102 \\ -.364 & -.373 & -.169 & -.207 & .378 & .387 \end{bmatrix},$$

and Equation 3.11, the standardized parameter estimates for the PCB data are

$$\hat{\mathbf{B}}^{*} = \begin{bmatrix} -.241 & -.223 & -.209 & -.278 & .170 & .118 \\ -.057 & .017 & .156 & .152 & .026 & -.069 \\ -.195 & -.207 & .004 & .016 & .257 & .292 \end{bmatrix}.$$

The standardized coefficients, based on variables scaled to mean zero and unit variance, admit to an interpretation of the relative magnitude, and possibly the relative importance, of the coefficients that cannot be achieved by the raw score regression coefficients. Examining the first row of $\hat{\mathbf{B}}^{*}$, it appears that age has roughly comparable effects on memory (immediate and delayed) and cognitive flexibility (Stroop color and word), with slightly less influence on the CVD risk factors of cholesterol and triglycerides. Gender appears to have greater influence on cognitive flexibility than either memory or CVD risk factors. Finally, the effect of PCBs adjusted for age and gender on both memory and CVD risk factors is noticeably greater than their effect on cognitive flexibility.

Column-wise interpretations also suggest certain overall patterns of relationships. Age and PCB exposure appear equally influential, and considerably more influential than gender, with respect to the memory and CVD risk factor variables, whereas age and gender appear to be the more important predictors of cognitive flexibility. It will become a matter of some interest in subsequent chapters to assess more formally if these apparent differences and patterns can be supported by evidence of statistical significance.

A Note on Computer Programs for Multivariate Linear Model Analysis

The computational details of multivariate linear model problems are sufficiently complex as to be nearly impossible without access to computer software. Computer programs for multivariate analysis are available from numerous commercial vendors and open source freeware providers. The analyses presented in this volume were analyzed by SPSS (MANOVA and GLM), SAS (PROC REG, PROC GLM, and PROC CANON), and STATA (MANOVA and CANON). For the majority of multivariate analyses, any of the above procedures will give comparable results, and we do not reproduce specific computer code here. In those instances where a computational option is unique and available in only one package, that procedure will be noted in the text.

Chapter Summary and a Look Ahead

Once a model has been chosen and specified, estimating the parameters of the population regression coefficients is the next step in conducting a multivariate general linear model analysis. These sample estimates allow one to begin to develop an understanding of how the predictor variables, collectively and uniquely, appear to be influencing the criterion variables in both magnitude and direction. Parameter estimation, however, is just a first step in the analytic process. In the two running examples with many standardized and unstandardized parameter estimates, it is easy to lose sight of the forest through the trees; one can begin to see how multivariate analyses can be useful in summarizing the relationships within **Y**, within **X**, and between **Y** and **X** and testing hypotheses about large amounts of information. A complete analysis of the relationship between sets of variables must begin with the specification of the model and the estimation of the parameters. The next stage of multivariate analysis focuses on quantifying the degree of association that exists between the sets of response and predictor variables in the model. To this end, we will develop multivariate measures of strength of association, R_m^2, that behave in a fashion similar to the univariate measure of R^2. Generalizing from univariate regression analysis, we will partition the joint variation among the set of response variables, **Y**, into its constituent parts—model and error—and assess the extent to which the joint variation in **Y** is predictable from the joint variation of the predictor variables, or subsets of the predictor variables, in **X**. In Chapter 4, we introduce the procedures for partitioning the multivariate sums of squares and cross-product matrices and developing four multivariate measures of strength of association.

CHAPTER 4. PARTITIONING THE SSCP, MEASURES OF STRENGTH OF ASSOCIATION, AND TEST STATISTICS

In univariate linear model analysis, the value of $R^2_{Y \cdot X_1 X_2 \ldots X_q}$ is a fundamental statistic for evaluating the extent to which variability in the response variable, Y, is accounted for as a function of the explanatory variables, $X_1, X_2, \ldots, X_q$. The value of $R^2_{Y \cdot X_1 X_2 \ldots X_q}$ is a useful scalar summary of the strength of relationship between response and explanatory variables and is defined as the ratio of the model sums of squares relative to the total sums of squares of Y. This ratio is based on the partition of the total sums of squares into its constituent sources of variability due to model and error,

$$SS_{TOTAL} = SS_{MODEL} + SS_{ERROR},$$

and in its most familiar form this measure of strength of association is given by the ratios,

$$R^2_{Y \cdot X_1 X_2 \ldots X_q} = \frac{SS_{MODEL}}{SS_{TOTAL}} = 1 - \frac{SS_{ERROR}}{SS_{TOTAL}} \qquad [4.1]$$

Defining measures of strength of association for the multivariate general linear model is a generalization of the univariate R^2 based on the partition of the multivariate sums of squares and cross-products (SSCP) *matrices* for total, model, and error terms of the partition. The multivariate partition of the SSCP matrices follows the same pattern,

$$\text{SSCP}_{TOTAL} = \text{SSCP}_{MODEL} + \text{SSCP}_{ERROR}.$$

From the partitioned SSCP matrices, it is possible to derive a scalar quantity from the ratio of $\text{SSCP}_{MODEL} / \text{SSCP}_{TOTAL}$ to compute a multivariate measure of strength of association, R^2_m, which carries the same conceptual meaning as in the univariate case (i.e., $0 < R^2_m < 1.00$). A definition of a multivariate R^2_m is complicated by the fact that there are four separate measures of strength of association commonly used in multivariate analysis.[1]

---

[1]The four multivariate measures of R^2_m and their associated test statistics include Wilks' Λ, Pillai's Trace, Hotelling's Trace, and Roy's GCR criterion. They differ largely in the way the arithmetic is performed on the partitioned SSCP matrices. Olson (1974, 1976) gives a comprehensive review of the relative advantages and disadvantages of each of the four. Cramer and Nicewander (1979) review these and other measures of strength of association.

Each of these four versions of R_m^2 is closely allied to its multivariate test statistic, just as the univariate R^2 is closely tied to its test statistic, F. In the sections to follow, we will (1) partition the mean corrected SSCP_{TOTAL} matrix into its constituent matrices SSCP_{MODEL} and SSCP_{ERROR} and (2) develop the definition of the four[2] commonly available measures of R_m^2.

Partition of the SSCP in the Multivariate General Linear Model

The Partition of Mean Corrected SSCP Matrices. A measure of the strength of association between the response variables in **Y** and the predictor variables in **X** for the multivariate model $\mathbf{Y} = \mathbf{XB} + \mathbf{E}$ depends on a partition of the matrix of total SSCP, $\text{SSCP}_{TOTAL} = \mathbf{Y}'\mathbf{Y}$, into SSCP matrices for both model and error. The model SSCP can be obtained by subtracting the SSCP_{ERROR}, defined in Equation 3.7, from $\mathbf{Y}'\mathbf{Y}$, giving the multivariate partition of the SSCP as

$$\mathbf{Y}'\mathbf{Y} = \hat{\mathbf{B}}'\mathbf{X}'\mathbf{Y} + (\mathbf{Y}'\mathbf{Y} - \hat{\mathbf{B}}'\mathbf{X}'\mathbf{Y}). \qquad [4.2]$$

The partition of Equation 4.2 is in uncorrected, raw score form, which was suitable for estimating the parameters of the model in Equation 3.4. For defining measures of multivariate strength of association and tests of hypotheses on models, it is more convenient to partition the SSCP matrices in mean corrected (centered) form. The mean corrected sum of squares for a single Y variable is $\text{SS}_Y = \Sigma Y^2 - \frac{1}{n}(\Sigma Y)^2$, where $\frac{1}{n}(\Sigma Y)^2 = n\bar{Y}^2$ is the mean correction factor to the uncorrected raw score ΣY^2. By analogy, the correction factor for an uncorrected, raw score SSCP matrix **Y** is $n\bar{\mathbf{Y}}'\bar{\mathbf{Y}}$, where $\bar{\mathbf{Y}}$ is an $n \times p$ matrix containing p columns of the means of the response

---

[2]Roy's GCR criterion cannot be directly computed as a ratio of SSCP matrices, but instead relies on the eigenvalues of the ratios of SSCP matrices. We introduce Roy's GCR here but delay a more complete discussion of eigenvalues and their relationship to multivariate test statistics until Chapter 7.

[3]The correction factor $n\bar{\mathbf{Y}}'\bar{\mathbf{Y}}$ is the SSCP matrix for the intercept terms of the model. Measures of strength of association and tests of hypotheses on the explanatory variables are based on the mean corrected SSCP matrices, which remove the intercepts from the tests. Leaving the intercept SSCP in the model would result in tests of hypotheses that constrain the intercepts to be zero. Hypotheses about the vector of intercepts in the regression model are possible but infrequently of interest. Separate partitions of the SSCP matrices for the intercept and for the extra SSCP due to the remaining predictors are described in Rencher (1998, p. 290) and Draper and Smith (1998, pp. 130–131).

variables.[3] Subtracting this correction factor from both sides of Equation 4.2 leads to the partition of mean corrected SSCP_{TOTAL} into its constituent parts,

$$(\mathbf{Y}'\mathbf{Y} - n\bar{\mathbf{Y}}'\bar{\mathbf{Y}}) = (\hat{\mathbf{B}}'\mathbf{X}'\mathbf{Y} - n\bar{\mathbf{Y}}'\bar{\mathbf{Y}}) + (\mathbf{Y}'\mathbf{Y} - \hat{\mathbf{B}}'\mathbf{X}'\mathbf{Y}),$$

$$\text{SSCP}_{TOTAL} = \text{SSCP}_{MODEL} + \text{SSCP}_{ERROR},$$

$$\mathbf{Q}_T = \mathbf{Q}_F + \mathbf{Q}_E. \qquad [4.3]$$

We denote the total and error SSCP matrices as $\mathbf{Q}_T$ and $\mathbf{Q}_E$, respectively, and the full model SSCP matrix consisting of the entire set of predictor variables for any given problem will be denoted by $\mathbf{Q}_F$. In later paragraphs, we will use $\mathbf{Q}_R$ to denote a restricted model SSCP matrix based on a subset of the predictors, and we will use $\mathbf{Q}_H$ to define the SSCP matrix associated with a specific hypothesis, often computed as the difference between full and restricted models.

The mean corrected SSCP matrices of Equation 4.3 can also be obtained from the mean corrected data matrices either by centering the data in **X** and **Y** or by applying appropriate correction factors to the raw score data SSCP matrices. The SSCP matrices within the explanatory variables, within the response variables, and the sums of cross products (SCP) between explanatory and response variables are found as

$$\mathbf{S}_{YY} = (\mathbf{Y}'\mathbf{Y} - n\bar{\mathbf{Y}}'\bar{\mathbf{Y}}),$$

$$\mathbf{S}_{YX} = (\mathbf{Y}'\mathbf{X} - n\bar{\mathbf{Y}}'\bar{\mathbf{X}}),$$

$$\mathbf{S}_{XY} = (\mathbf{X}'\mathbf{Y} - n\bar{\mathbf{X}}'\bar{\mathbf{Y}}),$$

$$\mathbf{S}_{XX} = (\mathbf{X}'\mathbf{X} - n\bar{\mathbf{X}}'\bar{\mathbf{X}}). \qquad [4.4]$$

These four building blocks (where $\mathbf{S}_{YY} = \mathbf{Q}_T$) of multivariate analysis are often collected into a single partitioned matrix **S** of all possible SSCP. The matrix is of order $(p+q \times p+q)$,

$$\mathbf{S}_{(p+q \times p+q)} = \begin{bmatrix} \mathbf{S}_{YY(p\times p)} & \mathbf{S}_{YX(p\times q)} \\ \mathbf{S}_{XY(q\times p)} & \mathbf{S}_{XX(q\times q)} \end{bmatrix}. \qquad [4.5]$$

Since these matrices are mean corrected, the least squares estimates of the parameters for the predictor variables, excluding the row vector of intercepts,[4] can be found by

$$\hat{\mathbf{B}}_{c(q\times p)} = \mathbf{S}_{XX}^{-1}\mathbf{S}_{XY}, \qquad [4.6]$$

which provides the same estimates of the parameters of the predictor variables of Equation 3.4 exclusive of the parameter estimates for X_0.

A partition of the mean corrected total SSCP matrix into model and error SSCP matrices equivalent to the partition defined in Equation 4.3 can be written in alternate form,

$$\mathbf{S}_{YY} = \hat{\mathbf{B}}_c\mathbf{S}_{XY} + (\mathbf{S}_{YY} - \hat{\mathbf{B}}_c\mathbf{S}_{XY}),$$

$$\mathbf{S}_{YY} = \mathbf{S}_{YX}\mathbf{S}_{XX}^{-1}\mathbf{S}_{XY} + (\mathbf{S}_{YY} - \mathbf{S}_{YX}\mathbf{S}_{XX}^{-1}\mathbf{S}_{XY}),$$

$$\mathbf{Q}_T = \mathbf{Q}_F + \mathbf{Q}_E. \qquad [4.7]$$

Substituting $\hat{\mathbf{B}}_c = \mathbf{S}_{XX}^{-1}\mathbf{S}_{XY}$ into the second line of Equation 4.7, the partition of $\mathbf{Q}_T$ into $\mathbf{Q}_F + \mathbf{Q}_E$ can also be expressed as function of the mean corrected SSCP of the **X** and **Y** variables, a result equal to that of the partition given in Equation 4.3.

The Partition of the SSCP Matrices for Standard Scores. Another useful partitioning of the SSCP of the linear model can be given in standard score terms. The model $\mathbf{Z}_Y = \mathbf{Z}_X\mathbf{B}^* + \mathbf{E}$ has SSCP_{TOTAL} defined by $\mathbf{Z}_Y'\mathbf{Z}_Y$, and from Equation 3.9 the SSCP_{ERROR} is $\mathbf{Z}_Y'\mathbf{Z}_Y - \mathbf{B}^{*\prime}\mathbf{Z}_X'\mathbf{Z}_Y$. By subtraction, the SSCP_{MODEL} is $\mathbf{B}^{*\prime}\mathbf{Z}'_X\mathbf{Z}_Y$. In standard score form, the bivariate correlation between the vectors Y and X is $r_{YX} = \frac{1}{n-1}Z_Y{}'Z_X$ and the $p\times q$ matrix of correlation coefficients between **Y** and **X** is found by $\mathbf{R}_{YX} = \frac{1}{n-1}\mathbf{Z}'_Y\mathbf{Z}_X$. Similarly, the remaining correlation matrices are found as $\mathbf{R}_{YY} = \frac{1}{n-1}\mathbf{Z}'_Y\mathbf{Z}_Y$, $\mathbf{R}_{XX} = \frac{1}{n-1}\mathbf{Z}'_X\mathbf{Z}_X$, and $\mathbf{R}_{XY} = \frac{1}{n-1}\mathbf{Z}'_X\mathbf{Z}_Y$, which can be collected into

---

[4]The row vector of intercepts, $\hat{\boldsymbol{\beta}}_{0k} = [\hat{\beta}_{01}, \hat{\beta}_{02}, \cdots \hat{\beta}_{0k}]$, can be recovered from $\bar{\mathbf{Y}} - \hat{\mathbf{B}}_c\bar{\mathbf{X}}$.

a single $p+q \times p+q$ correlation matrix with quadrants defined by one of the within-variable or between-variable correlation submatrices:

$$\mathbf{R}_{(p+q \times p+q)} = \begin{bmatrix} \mathbf{R}_{YY_{(p \times p)}} & \mathbf{R}_{YX_{(p \times q)}} \\ \mathbf{R}_{XY_{(q \times p)}} & R_{XX_{(q \times q)}} \end{bmatrix}. \qquad [4.8]$$

With the standardized regression coefficients estimated from correlation matrices, $\widehat{\mathbf{B}}^{*} = \mathbf{R}_{XX}^{-1}\mathbf{R}_{XY}$, the partitioning of the SSCP matrices for standard scores can be achieved as in Equation 4.9,

$$\mathbf{Z}_Y'\mathbf{Z}_Y = \widehat{\mathbf{B}}^{*\prime}\mathbf{Z}_X'\mathbf{Z}_Y + (\mathbf{Z}_Y'\mathbf{Z}_Y - \widehat{\mathbf{B}}^{*\prime}\mathbf{Z}_X'\mathbf{Z}_Y)$$

$$\mathbf{R}_{YY} = \widehat{\mathbf{B}}^{*\prime}\mathbf{R}_{XY} + (\mathbf{R}_{YY} - \widehat{\mathbf{B}}^{*\prime}\mathbf{R}_{XY})$$

$$\mathbf{R}_{YY} = \mathbf{R}_{YX}\mathbf{R}_{XX}^{-1}\mathbf{R}_{XY} + (\mathbf{R}_{YY} - \mathbf{R}_{YX}\mathbf{R}_{XX}^{-1}\mathbf{R}_{XY})$$

$$\mathbf{Q}_T^{*} = \mathbf{Q}_F^{*} + \mathbf{Q}_E^{*}. \qquad [4.9]$$

The Elements of the Partitioned SSCP Matrices. From Equation 4.7, the symbolic elements of the partition $\mathbf{Q}_T = \mathbf{Q}_F + \mathbf{Q}_E$ of the ($p \times p$) SSCP_{TOTAL} matrix into model and error SSCP matrices for a *p* variable problem would contain the elements displayed in this expanded version of the partitioning,

$$\begin{bmatrix} SS_{TOTAL_{Y_1}} & SCP_{TOTAL_{Y_1Y_2}} & \cdots & SCP_{TOTAL_{Y_1Y_p}} \\ SCP_{TOTAL_{Y_2Y_1}} & SS_{TOTAL_{Y_2}} & \cdots & SCP_{Y_1Y_2} \\ \vdots & \vdots & \ddots & \vdots \\ SCP_{TOTAL_{Y_pY_1}} & SCP_{TOTAL_{Y_pY_2}} & \cdots & SS_{TOTAL_{Y_p}} \end{bmatrix} =$$

$$\begin{bmatrix} SS_{MODEL_{Y_1}} & SCP_{MODEL_{Y_1Y_2}} & \cdots & SCP_{MODEL_{Y_1Y_p}} \\ SCP_{MODEL_{Y_2Y_1}} & SS_{MODEL_{Y_2}} & \cdots & SCP_{MODEL_{Y_2Y_p}} \\ \vdots & \vdots & \ddots & \vdots \\ SCP_{MODEL_{Y_pY_1}} & SCP_{MODEL_{Y_pY_2}} & \cdots & SS_{MODEL_{Y_p}} \end{bmatrix}$$

$$+ \begin{bmatrix} SS_{ERROR_{Y_1}} & SCP_{ERROR_{Y_1Y_2}} & \cdots & SCP_{ERROR_{Y_1Y_p}} \\ SCP_{ERROR_{Y_2Y_1}} & SS_{ERROR_{Y_2}} & \cdots & SCP_{ERROR_{Y_2Y_p}} \\ \vdots & \vdots & \ddots & \vdots \\ SCP_{ERROR_{Y_pY_1}} & SCP_{ERROR_{Y_pY_2}} & \cdots & SS_{ERROR_{Y_p}} \end{bmatrix}.$$

The quantities necessary to define the univariate values of $R^2_{Y_k \cdot X_1 X_2 \cdots X_q}$ can be retrieved from the main diagonal of these partitioned matrices. For each of the p response variables in the model, the proportion of variance in Y accounted for by $X_1, X_2, \cdots, X_q$ is found by the ratio $\mathrm{SSCP}_{MODEL} / \mathrm{SSCP}_{TOTAL}$. Element-wise division of the values on the main diagonal of $\mathbf{Q}_F$ (the univariate model SS of $Y_1, Y_2, \cdots, Y_p$) by the corresponding elements on the main diagonal of $\mathbf{Q}_T$ (the univariate total SS of $Y_1, Y_2, \cdots, Y_p$) give, respectively, the univariate full model $R^2_{Y_1 \cdot X_1 X_2 \cdots X_q}$, $R^2_{Y_2 \cdot X_1 X_2 \cdots X_q}, \ldots,$ $R^2_{Y_p \cdot X_1 X_2 \cdots X_q}$. These, and subsequent, univariate results are achieved as a by-product of the matrix-based solution to multivariate linear model problems.

The Elements of the Partitioned SSCP Matrices in Standard Score Form. Partitioning the SSCP matrices in standard score form by Equation 4.9 is an even more direct route to some useful univariate quantities. The symbolic partition $\mathbf{Q}^*_T = \mathbf{Q}^*_F + \mathbf{Q}^*_E$ is known to be

$$
\begin{bmatrix}
1.00 & R_{Y_1Y_2} & \cdots & R_{Y_1Y_p} \\
R_{Y_1Y_2} & 1.00 & \cdots & R_{Y_2Y_p} \\
\vdots & \vdots & \ddots & \vdots \\
R_{Y_pY_1} & R_{Y_pY_2} & \cdots & 1.00
\end{bmatrix}
=
\begin{bmatrix}
R^2_{Y_1 \cdot X_1 X_2 \cdots X_q} & & & \\
 & R^2_{Y_2 \cdot X_1 X_2 \cdots X_q} & & \\
 & & \ddots & \\
 & & & R^2_{Y_p \cdot X_1 X_2 \cdots X_q}
\end{bmatrix}
$$

$$
+
\begin{bmatrix}
1 - R^2_{Y_1 \cdot X_1 X_2 \cdots X_q} & & & \\
 & 1 - R^2_{Y_1 \cdot X_1 X_2 \cdots X_q} & & \\
 & & \ddots & \vdots \\
 & & & 1 - R^2_{Y_1 \cdot X_1 X_2 \cdots X_q}
\end{bmatrix}.
$$

The main diagonal elements of $\mathbf{Q}^*_T, \mathbf{Q}^*_F,$ and $\mathbf{Q}^*_E$ contain useful information. $\mathbf{Q}^*_T$ is the response variable correlation matrix $\mathbf{R}_{YY}$, which gives an idea of the multicollinearity among the p response variables that will require adjustment in subsequent steps of the analysis. The univariate R^2 s for each response variable appear on the main diagonal[5] of $\mathbf{Q}^*_F$ and the proportion of variation in each of the Y variables that is not accounted for by the q predictors $(1 - R^2)$ appears on the main diagonal of $\mathbf{Q}^*_E$.

---

[5]The off-diagonal elements of the standardized partitioned matrices have been left blank as they contain quantities that have no immediate interpretive value; they will play an important role in adjusting for the redundant variance within both **Y** and **X** since truly multivariate measures of strength of association, and their corresponding test statistics, must adjust for the correlations within variable sets while assessing the between-sets relationships.

Stewart and Love (1968) proposed a redundancy index (R^2_{dYX}, R^2_{dXY}) as a measure of the multivariate strength of association between **Y** and **X** that is a function of the diagonal elements of $\mathbf{Q}^*_F$. The redundancy index estimates the proportion of variance in one set of variables that is predictable from the variance in a second set of variables. The measure is asymmetric, that is, $R^2_{dYX} = R^2_{dXY}$ only if $p = q$. The redundancy index of the variance in **Y** predictable from **X** can be shown (Thompson, 1984, pp. 25–30) to be an additive function of the elements on the main diagonal of $\mathbf{Q}^*_F$,

$$R^2_{dYX} = \frac{1}{p} Tr\left[\mathbf{Q}^*_F\right] = \frac{1}{p} \Sigma R^2_{Y_k \bullet X_1 X_2 \cdots X_q}. \qquad [4.10]$$

The proportion of variance in **X** that is predictable from **Y** is obtained by reversing the standardized regression model to $\mathbf{Z_X} = \mathbf{B}^* \mathbf{Z_Y} + \mathbf{E}$, partitioning the SSCP matrices, and evaluating,

$$R^2_{dXY} = \frac{1}{q} Tr\left[\mathbf{Q}^*_F\right] = \Sigma R^2_{X_j \bullet Y_1 Y_2 \cdots Y_p}. \qquad [4.11]$$

The redundancy coefficient is not a truly multivariate index since it is simply the average of the univariate R^2s. Consequently, R^2_{dYX} or R^2_{dXY} will overestimate the proportion of joint variance in each set of variables predictable from its opposite set. The overestimation results from a failure to adjust for multicollinearity among the variables in **Y**; removing the redundancy among the response variables is a key feature of multivariate analysis.[6] The problem is visually illustrated in the Venn diagram of Figure 4.1, illustrating multicollinearity among the response variables, a condition that is usually encountered in practice.

An important goal of multivariate analysis is to adjust for the overlap among the response variables (areas a, b, and c of Figure 4.1) just as multicollinearity must be addressed among the predictor variables. The off-diagonal elements of $\mathbf{Q}^*_T$ are the correlations between the Y variables and give a sense of the extent of the multicollinearity and the degree of adjustment that would be required. Whatever the overlap between $Y_1, Y_2, \cdots, Y_p$, the redundant variance cannot be counted more than once in attempting to define a legitimate measure of joint variance for the Y set of variables. We

[6]The redundancy index is already adjusted for the collinearity among the X variables as part of the solution to the univariate R^2s.

Figure 4.1 Venn Diagram of the Overlap Among the Response Variables

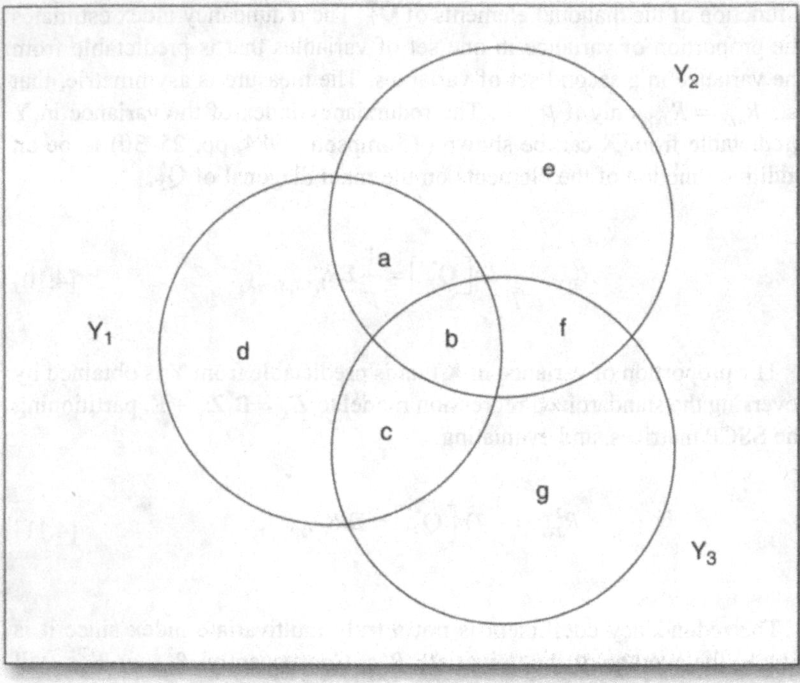

discuss alternative methods of adjusting for multicollinearity in a later section, but for the present it is sufficient to say that an important advantage of multivariate linear model analysis is the adjustment made for the redundant variance among both the response and the explanatory variables. The redundancy index is an *unadjusted* average of univariate R^2s and in practice will usually yield an overestimate of the shared variance between **Y** and **X**.[7] We introduce collinearity-adjusted, symmetric measures of multivariate association in the later sections of this chapter. Prior to that discussion, we illustrate the partitioning of the SSCP matrices by example with the personality data of Table 2.1 and the polychlorinated biphenyls (PCB) data of Table 2.2.

[7]The values of R^2_{dYX} or R^2_{dXY} will be correct estimates of the multivariate relationship only if the variables in **Y** are orthogonal.

Example 1: Personality and the Job Application Process

Evaluating the relationship between the multivariate set of $p = 4$ job interview variables (background preparation, social preparation, follow-up interviews, and job offers) and the $q = 3$ personality variables (Neuroticism, Extraversion, and Conscientiousness) requires that the joint variability in the job interview variables be partitioned into model and error sources of variance. From Equation 4.3, the partition of the 4×4 SSCP matrices for the personality data of Table 2.1 is summarized in Table 4.1.

Table 4.1 Partition of the 4 × 4 SSCP Matrices for the Personality Data

$$\text{Model SSCP } \mathbf{Q}_F = \begin{bmatrix} 166.727 & 36.425 & 14.292 & -2.200 \\ 36.425 & 378.341 & 29.068 & 22.087 \\ 14.292 & 29.068 & 3.352 & 1.008 \\ -2.200 & 22.087 & 1.008 & 1.881 \end{bmatrix}$$

$$\text{Error SSCP } \mathbf{Q}_E = \begin{bmatrix} 1496.764 & 808.080 & 22.046 & -17.584 \\ 808.080 & 2052.098 & 47.799 & 18.908 \\ 22.046 & 47.799 & 16.493 & 5.320 \\ -17.584 & 18.908 & 5.320 & 10.124 \end{bmatrix}$$

$$\text{Total SSCP } \mathbf{Q}_T = \begin{bmatrix} 1663.491 & 844.504 & 36.338 & -19.784 \\ 844.504 & 2430.439 & 76.866 & 40.995 \\ 36.338 & 76.866 & 19.845 & 6.328 \\ -19.784 & 40.995 & 6.328 & 12.005 \end{bmatrix}$$

The immediately useful information in Table 4.1 is the ratio of the model SS to the total SS on the main diagonals of $\mathbf{Q}_F$ and $\mathbf{Q}_T$ that yields the univariate whole model R^2 for each dependent variable predicted from the $q = 3$ predictors. The R^2s can be obtained by computing the ratio of SS_F/SS_T for corresponding elements on the main diagonals of $\mathbf{Q}_F$ and $\mathbf{Q}_T$. Thus for Y_1, $R^2_{\text{BACKGROUND}\cdot X_1X_2X_3X_4} = \dfrac{166.727}{1663.491} = .10$, and for the remaining three response variables, $R^2_{\text{social}\cdot X_1X_2X_3X_4} = .16$, $R^2_{\text{interviews}\cdot X_1X_2X_3X_4} = .17$, and $R^2_{\text{offers}\cdot X_1X_2X_3X_4} = .16$. These values can also be easily retrieved from the partition of the SSCP matrices in standard score form as defined in Equation 4.7. These partitioned matrices are summarized in Table 4.2.

Table 4.2 Partition of the 4 × 4 SSCP Matrices for the Personality Data in Standard Score Form

Model SSCP $\mathbf{Q}_F^* = \mathbf{R}_{YX}\mathbf{R}_{XX}^{-1}\mathbf{R}_{XY} =$	$\begin{bmatrix} .100 & .018 & .079 & -.016 \\ .018 & .156 & .132 & .129 \\ .079 & .132 & .169 & .065 \\ -.016 & .129 & .065 & .157 \end{bmatrix}$
Error SSCP $\mathbf{Q}_E^* = \mathbf{R}_{YY} - \mathbf{R}_{YX}\mathbf{R}_{XX}^{-1}\mathbf{R}_{XY} =$	$\begin{bmatrix} .900 & .402 & .121 & -.124 \\ .402 & .844 & .218 & .111 \\ .121 & .218 & .831 & .345 \\ -.124 & .111 & .345 & .843 \end{bmatrix}$
Total SSCP $\mathbf{Q}_T^* = \mathbf{R}_{YY} =$	$\begin{bmatrix} 1.000 & .420 & .200 & -.140 \\ .420 & 1.000 & .350 & .240 \\ .200 & .350 & 1.000 & .410 \\ -.140 & .240 & .410 & 1.000 \end{bmatrix}$

These univariate multiple R^2s, appearing on the main diagonal of $\mathbf{Q}_F^*$, have the usual interpretation; between 10% and 17% of the variance of the individual response variables is accounted for by the set of personality variables of Neuroticism, Extraversion, and Conscientiousness. While these univariate results are informative and will be used as follow-up tests to the multivariate evaluation, they do not themselves suggest a clear answer to the question of how much of the joint, nonredundant variance in **Y** is a function of the joint, nonredundant variance in **X**. As an upper limit, the joint variance in **Y** predictable from **X** could not exceed $R^2_{dYX} = \dfrac{.100+.156+.169+.157}{4} = .146$ and would require that the correlation matrix $\mathbf{R}_{YY}$ be an identity matrix. The nonorthogonality of the job application variables in **Y**, as illustrated in Figure 4.1, is documented in the off-diagonal elements of $\mathbf{Q}_T^*$. The off-diagonal elements of $\mathbf{Q}_E^*$ and $\mathbf{Q}_F^*$ will be used to adjust for this overlap in the definition of truly multivariate measures of association.

Example 2: The PCB Data

Partitioning the SSCP matrices for the multivariate linear model is illustrated with the $p = 6$ response variables of the PCB data measuring cognitive functioning (immediate and delayed visual memory, and the Stroop color and word tests) and cardiovascular disease risk factors (cholesterol and triglycerides). These response variables predicted from $q = 3$ explanatory variables of age, gender, and exposure to PCBs (log transformed) were introduced in Table 2.2. The mean corrected partition of the raw score SSCP matrices for the PCB data can be obtained from the estimates $\hat{\mathbf{B}}$ and Equation 4.3. The full model, error, and total SSCP matrices are summarized in Table 4.3. By themselves the individual elements of the partitioned raw score SSCP matrices have limited interpretive value but they are presented here for future reference.

Conversely, the elements of the partitioned standard score model are more immediately informative. Using the estimated parameters for the standard score model ($\hat{\mathbf{B}}^*$) of Chapter 3 and Equation 4.7, the partition of the standard score SSCP matrices in the correlation metric is summarized in Table 4.4.

Table 4.3 Partition of the 6 × 6 SSCP Matrices for the PCB Data in Mean Corrected Raw Score Form

$$\text{Model SSCP } \mathbf{Q}_F = \begin{bmatrix} .167 & .161 & .073 & .095 & -.160 & -.149 \\ .161 & .161 & .082 & .103 & -.158 & -.155 \\ .073 & .082 & .064 & .075 & -.074 & -.083 \\ .095 & .103 & .075 & .090 & -.094 & -.100 \\ -.160 & -.158 & -.074 & -.094 & .158 & .152 \\ -.149 & -.155 & -.083 & -.100 & .152 & .158 \end{bmatrix}$$

$$\text{Error SSCP } \mathbf{Q}_E = \begin{bmatrix} .833 & .618 & .129 & .076 & .046 & .079 \\ .618 & .869 & .127 & .090 & .016 & .055 \\ .129 & .127 & .936 & .659 & -.031 & .039 \\ .076 & .090 & .659 & .910 & -.049 & .021 \\ .046 & .016 & -.031 & -.049 & .842 & .409 \\ .079 & .055 & .039 & .021 & .409 & .842 \end{bmatrix}$$

$$\text{Total SSCP } \mathbf{Q}_T = \begin{bmatrix} 1.000 & .779 & .202 & .172 & -.114 & -.070 \\ .779 & 1.000 & .209 & .193 & -.142 & -.100 \\ .202 & .209 & 1.000 & .733 & -.104 & -.044 \\ .172 & .193 & .733 & 1.000 & -.143 & -.080 \\ -.114 & -.142 & -.104 & -.143 & 1.000 & .561 \\ -.070 & -.100 & -.044 & -.080 & .561 & 1.000 \end{bmatrix}$$

Table 4.4 Partition of the 6 × 6 SSCP Matrices for the PCB Data in Standard Score Form

Model SSCP

$$\mathbf{Q}_F^* = \mathbf{R}_{YX}\mathbf{R}_{XX}^{-1}\mathbf{R}_{XY} = \begin{bmatrix} .167 & .161 & .073 & .095 & -.160 & -.149 \\ .161 & .161 & .082 & .103 & -.158 & -.155 \\ .073 & .082 & .064 & .075 & -.074 & -.083 \\ .095 & .103 & .075 & .090 & -.094 & -.100 \\ -.160 & -.158 & -.074 & -.094 & .158 & .152 \\ -.149 & -.155 & -.083 & -.100 & .152 & .158 \end{bmatrix}$$

Error SSCP

$$\mathbf{Q}_E^* = \mathbf{R}_{YY} - \mathbf{R}_{YX}\mathbf{R}_{XX}^{-1}\mathbf{R}_{XY} = \begin{bmatrix} .833 & .618 & .129 & .076 & .046 & .079 \\ .618 & .869 & .127 & .090 & .016 & .055 \\ .129 & .127 & .936 & .659 & -.031 & .039 \\ .076 & .090 & .659 & .910 & -.049 & .021 \\ .046 & .016 & -.031 & -.049 & .842 & .409 \\ .079 & .055 & .039 & .021 & .409 & .842 \end{bmatrix}$$

Total SSCP

$$\mathbf{Q}_T^* = \mathbf{R}_{YY} = \begin{bmatrix} 1.000 & .779 & .202 & .172 & -.114 & -.070 \\ .779 & 1.000 & .209 & .193 & -.142 & -.100 \\ .202 & .209 & 1.000 & .733 & -.104 & -.044 \\ .172 & .193 & .733 & 1.000 & -.143 & -.080 \\ -.114 & -.142 & -.104 & -.143 & 1.000 & .561 \\ -.070 & -.100 & -.044 & -.080 & .561 & 1.000 \end{bmatrix}$$

The diagonal elements of $\mathbf{Q}_F^*$ contain the univariate R^2s of the six dependent variables predicted from the three predictors of age, gender, and PCB exposure. About 16% to 17% of the variability in the memory variables is predictable from the collection of the predictors, whereas about 6% to 9% of the variability in the cognitive flexibility measures and about 16% of the variability in the cardiovascular risk factors are accounted for by the predictor set of age, gender, and PCB body burden. Although these percentages of variance accounted for could be considered "medium to large" by some standards (Cohen, 1988), it is not yet certain that any of the effects exceed their sampling errors. Moreover, the magnitude of the univariate R^2s is not a faithful reflection of a multivariate measure of variance accounted for unless the off-diagonal elements $\mathbf{R}_{YY}$ are zero. Assuming $\mathbf{R}_{YY} = \mathbf{I}$, an upper limit of the joint variance in **Y** predictable from **X** would be estimated as $R^2_{dYX} = \frac{.167 + .161 + .064 + .090 + .158 + .158}{6} = .133$. The off-diagonal elements of $\mathbf{Q}^*_T = \mathbf{R}_{YY}$ for these data are clearly nonzero, in some instances substantially so (e.g., .779, .733, and .561), which suggests that the redundancy index would overestimate the relationship between **Y** and **X**. We introduce better multivariate measures of association in the following sections.

Further Partitioning of the SSCP Matrices: Full and Restricted Models and Defining $\mathbf{Q}_H$

Tests of hypotheses on the regression coefficients in linear model analysis can be conceptualized as tests of differences in proportions of variance accounted for between model specifications (Rindskopf, 1984). Every hypothesis implies a restriction to be placed on the parameters of the full model; imposing such a restriction on the full model leads to the estimation of a restricted model consistent with the hypothesis. Every multivariate full and restricted model will possess a dedicated measure of goodness of fit of model to data, that is, a multivariate measure strength of association, $R^2_{m_F}$ and $R^2_{m_R}$. If the hypothesis is false, then these two measures of goodness of fit will diverge in magnitude. A test statistic can be employed to evaluate the statistical significance of this divergence. One view of hypothesis testing involves evaluating differences between measures of goodness of fit of full and restricted models, each based on a separate partitioning of the SSCP. When applied to a multivariate linear model, this strategy is a generalization of the univariate extra sums of squares approach discussed in Chapter 1 (Draper & Smith, 1998, Chap. 6). Generalizations of this extra sums of squares strategy to the multivariate case is given in Rencher (2002, pp. 330–331).

By convention, let the initial specification of the model **Y** = **XB** + **E** be a definition of the linear model containing all the predictor variables of interest, and let the total SSCP matrix of this model be denoted by $\mathbf{Q}_T$ as defined in the previous paragraphs. Also by convention, let the model SSCP matrix defined by $\hat{\mathbf{B}}\mathbf{X}'\mathbf{Y} - n\bar{\mathbf{Y}}'\bar{\mathbf{Y}} = \mathbf{Q}_F$ be the *full model* SSCP that is attributable to the *full model design matrix*, **X**,[8] and let the error SSCP matrix, $\mathbf{Q}_E = \mathbf{Q}_T - \mathbf{Q}_F$, denote the full model error SSCP. Once a hypothesis has

[8]Different textbooks and software manuals label the full model as the *whole model* or the *overall model*. The terms are synonymous and define the full set of predictor variables specified by the data analyst.

been formulated on some substantive basis, the restrictions implied in that hypothesis can be imposed on the specification of the full model to define a *restricted model*.[9] Imposing these restrictions on the full model has the effect of deleting designated predictors from the full model design matrix and creating a restricted model design matrix, *say* $\mathbf{X}_R$. The parameters of the restricted model, say $\mathbf{B}_R$, may now be obtained from the restricted model design matrix and can be used to define a new, reduced model that is consistent with the specified hypothesis,

$$\mathbf{Y} = \mathbf{X_R B_R} + \mathbf{E}. \qquad [4.12]$$

To illustrate the full-to-restricted model strategy, assume a $p=2, q=2+1$. multivariate linear model with full model unstandardized parameter matrix **B** (including rows for the intercept and two predictors) given by

$$\mathbf{B} = \begin{bmatrix} \beta_{01} & \beta_{02} \\ \beta_{11} & \beta_{12} \\ \beta_{21} & \beta_{22} \end{bmatrix}.$$

A null hypothesis that stipulates that the substantive predictor variables in **X** have no relationship to the response variables in **Y** would be stated by setting the values of the four population regression coefficients in **B**, exclusive of the intercept, simultaneously equal to zero. That is,

$$H_0: \begin{bmatrix} \beta_{11} & \beta_{12} \\ \beta_{21} & \beta_{22} \end{bmatrix} = \begin{bmatrix} 0 & 0 \\ 0 & 0 \end{bmatrix}.$$

This hypothesis implies a restriction to be placed on the full model matrix **B** and leads to the definition of a restricted model specification of Equation 4.12 of

$$\mathbf{B}_R = \begin{bmatrix} \beta_{01} & \beta_{02} \\ 0 & 0 \\ 0 & 0 \end{bmatrix} = \begin{bmatrix} \beta_{01} & \beta_{02} \end{bmatrix}.$$

---

[9]Restricting the full model can take many different forms. For the moment, we consider only those restrictions that delete predictor variables from the full model design matrix.

Imposing the restriction implied in H_0 on the full model specifies a *restricted model design matrix* $\mathbf{X}_R$ that deletes X_1 and X_2 and that includes only the intercept column vector X_0.

From Equation 4.3, the full model SSCP matrix is evaluated as $\mathbf{Q}_F = \hat{\mathbf{B}}'\mathbf{X}'\mathbf{Y} - n\bar{\mathbf{Y}}'\bar{\mathbf{Y}}$. The parameters of the restricted model are estimated by substituting $\hat{\mathbf{B}}_R$ for $\mathbf{B}_R$ in Equation 4.13:

$$\hat{\mathbf{B}}_R = (\mathbf{X}_R'\mathbf{X}_R)^{-1}\mathbf{X}_R'\mathbf{Y} \qquad [4.13]$$

and the restricted model SSCP is obtained by $\mathbf{Q}_R = \hat{\mathbf{B}}_R\mathbf{X}_R'\mathbf{Y} - n\bar{\mathbf{Y}}'\bar{\mathbf{Y}}$. The hypothesis SSCP matrix, $\mathbf{Q}_H$, is obtained as a difference between full and restricted model SSCP matrices,

$$\mathbf{Q}_H = \mathbf{Q}_F - \mathbf{Q}_R = \hat{\mathbf{B}}'\mathbf{X}'\mathbf{Y} - \hat{\mathbf{B}}_R\mathbf{X}_R'\mathbf{Y}. \qquad [4.14]$$

The hypothesis SSCP matrix $\mathbf{Q}_H$ defines the incremental influence of the variables that were deleted from the full model over and above those variables that were left in the restricted model—in the case of this example, the intercept terms alone were retained in $\mathbf{X}_R$.

Hypotheses on the individual predictor variables can be similarly constructed. Testing the hypothesis that the predictor X_1 adjusted for X_2 equals zero would be stated as

$$H_0\colon [\beta_{11} \quad \beta_{12}] = [0 \quad 0].$$

Placing this restriction on the full model would imply the parameters of the restricted model B_R as

$$\mathbf{B}_R = \begin{bmatrix} \beta_{01} & \beta_{02} \\ 0 & 0 \\ \beta_{21} & \beta_{22} \end{bmatrix} = \begin{bmatrix} \beta_{01} & \beta_{02} \\ \beta_{21} & \beta_{22} \end{bmatrix}.$$

To maintain the conformability requirement of matrix multiplication, this restricted model parameter matrix requires that the variable X_1 be deleted from the full model design matrix, thus defining the restricted model design matrix, $\mathbf{X}_R$. Substituting the least squares estimates of the parameters, $\hat{\mathbf{B}}_R$, into Equation 4.13 and evaluating Equation 4.14 defines the hypothesis SSCP matrix consistent with the hypothesis that variable X_1 adds nothing to the prediction of $\mathbf{Y}$ beyond that which can already be accounted for by X_2. Testing the hypothesis that X_2 adjusted for X_1 adds nothing to the multivariate prediction of $\mathbf{Y}$ that beyond X_1 would proceed similarly; the hypothesis would state that the third row of $\mathbf{B}$ for the full model be set to zero, with the consequence of dropping the variable X_2 from the full model design matrix followed by an estimate of $\mathbf{Q}_R$ and $\mathbf{Q}_H$ for this hypothesis. In general, a test of hypothesis on any subset of predictor variables in a multivariate model, say X_1 and X_2 adjusted for $X_3, X_4, \ldots, X_q$, can be constructed in this fashion, as can tests of more complicated hypotheses.

The *extra sums of squares* based on differences between full and restricted models in the multivariate case is analogous to the same procedure that was used to define the squared partial and semipartial correlations of Chapter 1. Placing restrictions on the multivariate full model consistent with different hypotheses is related to a variety of multivariate measures of association and their tests of significance, including tests of the full model regression ($R^2_{\mathbf{Y}\cdot\mathbf{X}}$), tests of partial or semipartial effects of individual predictors ($R^2_{\mathbf{Y}\cdot X_1|X_2X_3\cdots X_q}$ and $R^2_{\mathbf{Y}\cdot(X_1|X_2X_3\cdots X_q)}$), and tests of partial or semipartial effects of sets of predictors $\left(R^2_{\mathbf{Y}\cdot\mathbf{X}_{set_1}|\mathbf{X}_{set_2}} \text{ and } R^2_{\mathbf{Y}(\mathbf{X}_{set_1}|\mathbf{X}_{set_2})}\right)$. To develop multivariate analogues of these measures, we generalize from the definitions of these quantities in the univariate case, and do so by using the SSCP matrices defined by $\mathbf{Q}_T, \mathbf{Q}_F, \mathbf{Q}_R, \mathbf{Q}_H$, and $\mathbf{Q}_E$.

Conceptual Definition of Some Multivariate Measures of Association

Multivariate measures of association conform conceptually to a pattern that is similar to their univariate counterparts. Consider the Venn diagram of Figure 4.2.[10]

Let the diagrammatic equivalent of the multivariate SSCP_{TOTAL} ($\mathbf{Q}_T$) be defined by the sum of all the areas within the boundaries of Y_1 and Y_2, that is, $\mathbf{Q}_T \equiv a + b + c + \cdots + m$. Define the portion of this SSCP_{TOTAL} that is held in common with all the overlapping areas of X_1, X_2, and X_3 (e.g., $\mathbf{Q}_F \equiv a + b + c + \cdots + j$), and define the portion of the $\text{SSCP}_{\text{TOTAL}}$ that is not

[10]The Venn diagram is a useful heuristic device for visualizing the concepts of overlap among multivariate sets of variables. It is metaphorical in the sense that it cannot accurately display negative correlations, suppressor variables, or enhancer variables. The symbols in Equations 4.15 to 4.18 are placed in quotes (" ") to emphasize that the definitions are conceptual and not computational. All the quantities in these equations are $p \times p$ matrices. Additional methods will be used to reduce the matrices to scalars with interpretable meanings as measures of multivariate R^2.

Figure 4.2 Venn Diagram of a $p = 2$, $q = 3$ Multivariate Regression Model

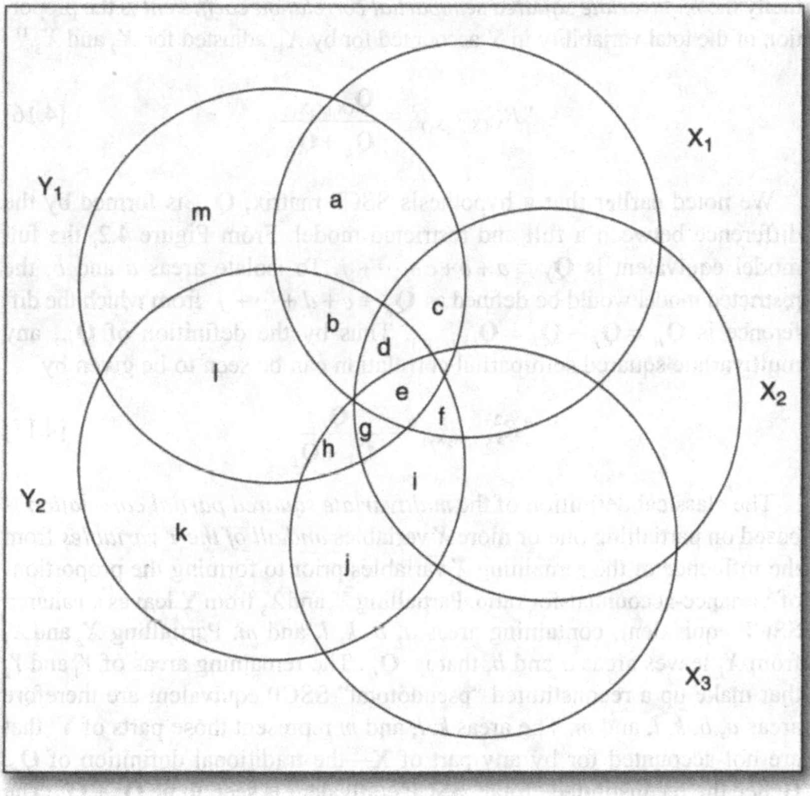

held in common with X_1, X_2, and X_3 as $\mathbf{Q}_E$. The conceptual definition of a *whole model multivariate squared correlation coefficient* is then given by

$$"R^2_{\mathbf{Y}\cdot\mathbf{X}}" = \frac{\mathbf{Q}_F}{\mathbf{Q}_T} = \frac{\mathbf{Q}_F}{\mathbf{Q}_E + \mathbf{Q}_F}, \qquad [4.15]$$

noting that $\mathbf{Q}_T = \mathbf{Q}_E + \mathbf{Q}_F$.

The classic definition of the squared *semipartial* correlation coefficient depends on partialling one or more X variables from the remaining X variables, and forming the ratio of this semipartialled X variability to the total variability in **Y**. Say that we wish to partial from X_1 those parts of X_1 that overlap with Y_1 or Y_2 and that are predictable from X_2 and X_3—that is, areas *c*, *d*, and *e* of the

Venn diagram. This partialling leaves the areas *a* and *b* of Figure 4.2 that are unique to X_1. If we let this partialled effect be denoted as $\mathbf{Q}_{(X_1|X_2X_3)}$, then conceptually the *multivariate squared semipartial correlation coefficient* is the proportion of the total variability in **Y** accounted for by X_1, adjusted for X_2 and X_3.[11]

$$"R^2_{\mathrm{Y}\cdot(\mathrm{X}_1|\mathrm{X}_2\mathrm{X}_3)}" = \frac{\mathbf{Q}_{(\mathrm{X}_1|\mathrm{X}_2\mathrm{X}_3)}}{\mathbf{Q}_E + \mathbf{Q}_F}. \qquad [4.16]$$

We noted earlier that a hypothesis SSCP matrix, $\mathbf{Q}_H$, is formed by the difference between a full and restricted model. From Figure 4.2, the full model equivalent is $\mathbf{Q}_F \equiv a+b+c+\cdots+j$. To isolate areas *a* and *b*, the restricted model would be defined as $\mathbf{Q}_R \equiv c+d+\cdots+j$ from which the difference is $\mathbf{Q}_H = \mathbf{Q}_F - \mathbf{Q}_R = \mathbf{Q}_{(X_1|X_2X_3)}$. Thus by the definition of $\mathbf{Q}_H$, any multivariate squared semipartial correlation can be seen to be given by

$$"R^2_{\mathrm{Y}\cdot(\mathrm{X}_1|\mathrm{X}_2\mathrm{X}_3)}" = \frac{\mathbf{Q}_H}{\mathbf{Q}_F + \mathbf{Q}_E}. \qquad [4.17]$$

The classical definition of the *multivariate squared partial correlation* is based on partialling one or more *X* variables *and all of the Y variables* from the influence of the remaining *X* variables prior to forming the proportion-of-variance-accounted-for ratio. Partialling X_2 and X_3 from **Y** leaves an altered SSCP equivalent, containing areas *a*, *b*, *k*, *l*, and *m*. Partialling X_2 and X_3 from X_1 leaves areas *a* and *b*, that is, $\mathbf{Q}_H$. The remaining areas of Y_1 and Y_2 that make up a reconstituted "pseudototal" SSCP equivalent are therefore areas *a*, *b*, *k*, *l*, and *m*. The areas *k*, *l*, and *m* represent those parts of **Y** that are not accounted for by any part of **X**—the traditional definition of $\mathbf{Q}_E$. Hence the reconstituted "total" SSCP equivalent is seen to be $\mathbf{Q}_H + \mathbf{Q}_E$. The multivariate squared partial correlation is therefore given as

$$"R^2_{(\mathrm{Y}|\mathrm{X}_2\mathrm{X}_3)\cdot(\mathrm{X}_1|\mathrm{X}_2\mathrm{X}_3)}" = \frac{\mathbf{Q}_H}{\mathbf{Q}_H + \mathbf{Q}_E}. \qquad [4.18]$$

Following these conceptual definitions, it is easy to see the connection between the SSCP matrices, $\mathbf{Q}_F, \mathbf{Q}_R, \mathbf{Q}_H$, and $\mathbf{Q}_E$, and these various measures of strength of association. Several such measures have evolved in the literature of multivariate analysis, all of which depend first on a partition of the SSCP, $\mathbf{Q}_T = \mathbf{Q}_F + \mathbf{Q}_E$, and second on the definition of $\mathbf{Q}_H$ resulting from

---

[11]The squared semipartial correlation must be a scalar value with limits $0 \le R^2_{\mathrm{Y}\cdot(\mathrm{X}_1|\mathrm{X}_2\mathrm{X}_3\cdots\mathrm{X}_q)} \le 1$. The definitions given in Equations 4.15 to 4.18 include matrices in numerator and denominator and are strictly conceptual; the computational methods for reducing $R^2_{\mathrm{Y}\cdot(\mathrm{X}_1|\mathrm{X}_2\mathrm{X}_3\cdots\mathrm{X}_q)}$ to a scalar are introduced below.

a statement of hypotheses and restrictions placed on full models consistent with these hypotheses.[12]

An Asymmetric Multivariate Measure of $\mathbf{R}^2$: Hooper's Squared Trace Correlation

There are several ways to define the proportion of variance in a multivariate *set* of response of variables (**Y**) that is shared with a *set* (whole or partial) of explanatory variables (**X**). These various definitions are often tied to a particular form of arithmetic associated with a corresponding multivariate test statistic (Cramer & Nicewander, 1979; Hooper, 1959). Four of these several measures are in common usage and will be discussed in the subsequent sections. Each of these multivariate measures of association can be best understood as a generalization, in one form or another, of the univariate R^2 of Equation 1.16. A first approximation to such a generalized measure is the ratio of model-to-total SSCP matrices,

$$"R^2_{\mathbf{Y}\cdot\mathbf{X}}" = \frac{\boldsymbol{SSCP}_{MODEL}}{\boldsymbol{SSCP}_{TOTAL}} = \mathbf{Q}_T^{*-1}\mathbf{Q}_F^{*}, \qquad [4.19]$$

which can be written in raw score form or standard score form as in Equation 4.19.[13] From the partition of the standard score SSCP matrices of Equation 4.9, recall that $\mathbf{Q}_T^{*} = \mathbf{R}_{YY}$ and $\mathbf{Q}_F = \mathbf{R}_{YX}\mathbf{R}_{XX}^{-1}\mathbf{R}_{XY}$, such that the ratio in Equation 4.19 can be expressed as a $p \times p$ quadruple product of four correlation matrices,

$$\mathbf{R}_{YY}^{-1}\mathbf{R}_{YX}\mathbf{R}_{XX}^{-1}\mathbf{R}_{XY}. \qquad [4.20]$$

The decomposition of this quantity into a scalar measure of association can be illustrated by examining its properties for a bivariate correlation, a multiple correlation, and a p-variable multivariate correlation. Assume for a moment that the problem is a bivariate regression model with $p = 1$ and $q = 1$ in which case $\mathbf{R}_{YY}^{-1} = r_{YY}^{-1} = 1$, $\mathbf{R}_{XX}^{-1} = r_{XX}^{-1} = 1$, and $r_{YX} = r_{XY}$. Under these

---

[12]If all the variables in the full model are to be tested, then $\mathbf{Q}_F = \mathbf{Q}_H$ and Equations 4.17 and 4.18 produce identical results. Conversely, if a subset of predictors or a single predictor variable are involved in the hypothesis and $\mathbf{Q}_F \neq \mathbf{Q}_H$, then the distinction between the squared semipartial and the squared partial correlation rests in the definition of the denominator of the measure of association as seen in Equations 4.17 and 4.18. All commercially available software for multivariate analysis automatically defaults to the definition of the squared partial correlation of Equation 4.18.

[13]The values of multivariate measures of R^2 are invariant to linear transformations of the variables. Raw score and standard score solutions yield the same values.

conditions, Equation 4.21 reduces to r^2_{YX}—the proportion of variance in Y accounted for by X. Consider further a univariate multiple regression model with $p = 1$ and $q > 1$, and then $\mathbf{R}^{-1}_{YY} = r^{-1}_{YY} = 1$, $\hat{\mathbf{B}}^{*} = \mathbf{R}^{-1}_{XX}\mathbf{r}_{XY}$, and Equation 4.20 defines $R^2_{Y \cdot X_1 X_2 \cdots X_q} = \mathbf{r}^{-1}_{YY}\mathbf{r}_{YX}\mathbf{R}^{-1}_{XX}\mathbf{r}_{XY}$—the proportion of variance in Y accounted for by $X_1, X_2, \cdots, X_q$.

For problems with $p > 1$ response variables and $q \geq 1$ predictor variables, Equation 4.20 is a first approximation to the multivariate generalization of proportion of variance in **Y** accounted for by **X**. This quantity alone, however, is a matrix of order $(p \times p)$ and not a scalar measure with lower and upper bounds of 0 and 1. Hooper (1959) provided a solution to this problem by taking the arithmetic mean of the trace of Equation 4.21. Hooper's Index of Trace Correlation, $\bar{r}^2$, is given by

$$\bar{r}^2 = \frac{1}{p} Tr\left[\mathbf{R}^{-1}_{YY}\mathbf{R}_{YX}\mathbf{R}^{-1}_{XX}\mathbf{R}_{XY}\right]. \qquad [4.21]$$

The value of $\bar{r}^2$ is the computational equivalent of the ratio

$$\frac{(a+b+\ldots+j)}{(a+b+\ldots+j+\ldots+m)}$$

of Figure 4.2. Equation 4.21, bounded by the interval [0 1], is interpreted as the proportion of the *joint, nonredundant variance in the Y set* of variables (i.e., Hooper's generalized variance) that is accounted for by the *joint, nonredundant variance in the X set* of variables. Some insight into the meaning of $\bar{r}^2$ can be achieved by a contrived example. Consider a situation where the variables in **Y** are mutually orthogonal ($\mathbf{R}_{YY} = \mathbf{I}$). Under these conditions, the trace of $\mathbf{R}^{-1}_{YY}\mathbf{R}_{YX}\mathbf{R}^{-1}_{XX}\mathbf{R}_{XY}$ in symbolic terms is

$$\begin{bmatrix} 1 & 0 \\ 0 & 1 \end{bmatrix} \begin{bmatrix} r_{Y_1X_1} & r_{Y_1X_2} \\ r_{Y_2X_1} & r_{Y_2X_2} \end{bmatrix} \begin{bmatrix} \dfrac{1}{1-r^2_{X_1X_2}} & \dfrac{-r_{X_1X_2}}{1-r^2_{X_1X_2}} \\ \dfrac{1-r_{X_1X_2}}{1-r^2_{X_1X_2}} & \dfrac{1}{1-r^2_{X_1X_2}} \end{bmatrix} \begin{bmatrix} r_{Y_1X_1} & r_{Y_2X_1} \\ r_{Y_1X_2} & r_{Y_2X_2} \end{bmatrix}. \qquad [4.22]$$

Performing the multiplications yields the diagonal elements

$$\begin{bmatrix} \dfrac{r^2_{Y_1 \cdot X_1} + r^2_{Y_1 \cdot X_1} - 2r_{Y_1X_1}r_{Y_1X_2}r_{X_1X_2}}{1-r^2_{X_1X_2}} & \\ & \dfrac{r^2_{Y_2 \cdot X_1} + r^2_{Y_2 \cdot X_1} - 2r_{Y_2X_1}r_{Y_2X_2}r_{X_1X_2}}{1-r^2_{X_1X_2}} \end{bmatrix} = \begin{bmatrix} R^2_{Y_1 \cdot X_1X_2} & \\ & R^2_{Y_2 \cdot X_1X_2} \end{bmatrix}.$$

Taking the arithmetic average of the trace of Equation 4.22 for $p = 2$ gives Hooper's Index,

$$\bar{r}^2 = \frac{1}{2}\left(R^2_{Y_1 \cdot X_1 X_2} + R^2_{Y_2 \cdot X_1 X_2}\right). \qquad [4.23]$$

Assuming orthogonal variables in **Y**, Hooper's $\bar{r}^2$ is therefore the arithmetic average of the p squared multiple correlations of each *orthogonalized* Y variables with **X**. Although it is not immediately apparent from this simplified example, the purpose of $\mathbf{R}_{YY}^{-1}$ and $\mathbf{R}_{XX}^{-1}$ in Equation 4.21 is to adjust for (i.e., orthogonalize) the variables within **Y** and within **X**. In a later section of this chapter, we give a further tutorial on an analogous process of adjusting for the redundancies in the Y variables in multivariate linear models.

Examples: Hooper's $\bar{r}^2$ for the Personality Data and the PCB Data

Hooper's $\bar{r}^2$ is estimated for the $q = 3$ personality predictor variables and the $p = 4$ job interview response variables of Table 2.1. From the partitioned SSCP matrices of Table 4.2, we evaluate $Tr(\mathbf{R}_{YY}^{-1}\mathbf{R}_{YX}\mathbf{R}_{XX}^{-1}\mathbf{R}_{XY})/p$, and using Equation 4.21 we find

$$\bar{r}^2 = \frac{1}{4}Tr\begin{pmatrix} \mathbf{.0995} & -.0433 & .0161 & -.0455 \\ -.0476 & \mathbf{.1349} & .0781 & .1252 \\ .0858 & .0680 & \mathbf{.1422} & -.0225 \\ -.0254 & .0630 & -.0095 & \mathbf{.1295} \end{pmatrix} = .169.$$

We conclude that about 17% of the nonredundant joint variance of the four job interview outcome variables is accounted for by the joint variance in Extraversion, Neuroticism, and Conscientiousness.

Applying Equation 4.21 to the $p = 6$, $q = 3$ variables of the PCB data gives

$$\bar{r}^2 = \frac{1}{6}Tr\begin{pmatrix} \mathbf{.1042} & .0887 & .0203 & .0353 & -.0932 & -.0695 \\ .0508 & \mathbf{.0610} & .0456 & .0511 & -.0566 & -.0695 \\ -.0154 & -.0068 & \mathbf{.0110} & .0085 & .0104 & -.0017 \\ .0598 & .0624 & .0463 & \mathbf{.0571} & -.0556 & -.0558 \\ -.0878 & -.0788 & -.0236 & -.0352 & \mathbf{.0819} & .0681 \\ -.0878 & -.0937 & -.0595 & -.0682 & .0901 & \mathbf{.1036} \end{pmatrix} = .070.$$

Approximately 7% of the nonredundant joint variance of the cardiovascular risk factors, memory variables, and cognitive flexibility variables is accounted for by the set of the predictors of age, gender, and PCB exposure.

Although Hooper's $\bar{r}^2$ is a pedagogically useful method for assessing the degree of overlap of **Y** and **X**, it has the disadvantage of being an asymmetric measure of association that can take on different values depending on the direction of the model, the number of predictors (q), and the number of criterion variables (p). The value of $\bar{r}^2$ will (possibly) be different for the model **Y** = **XB** + **E** than for the reciprocal model **X** = **YB** + **E** if the number of criterion and predictor variables in the model are not equal. Thus, there are two asymmetric measures of Hooper's Trace, $\bar{r}^2_{Y \cdot X}$ and $\bar{r}^2_{X \cdot Y}$, which will agree only if $p = q$. For the **X** = **YB** + **E** model applied to the personality data, we find $\bar{r}^2_{X \cdot Y} = .127$ (vs. .169), and for the PCB data the comparable value is $\bar{r}^2_{X \cdot Y} = .140$ (vs. .070). Unlike the redundancy index discussed in earlier sections, Hooper's Trace adjusts for the confounding between the Y variables and between the X variables, but if $p \neq q$ the index is asymmetric. To avoid these difficulties, several symmetric measures of association have been proposed (Cramer & Nicewander, 1979). Four symmetric measures of multivariate R^2_m are commonly employed in multivariate linear model analysis. Three of these measures—Pillai's Trace V (Pillai, 1955), Wilks' Λ (Wilks, 1932), and the Lawley-Hotelling Trace T (Hotelling, 1951; Lawley, 1938)—follow closely on the logic of Hooper's index of trace correlation but differ in their arithmetic approach. The fourth measure, Roy's (1957) greatest characteristic root (GCR), depends on the solution of the eigenvalue problem and will be briefly introduced here and given more extensive attention in Chapter 7.

Relationships Between the Univariate and Multivariate R²s and Their Test Statistics

Every multivariate test statistic (V, Λ, T, and θ) is directly connected to a measure of strength of association (R^2_V, R^2_Λ, R^2_T, and R^2_θ) in the same way that R^2 and the F-test are related in univariate regression analysis. The purpose of the conceptual definition of multivariate measures of association of Figure 4.2 was to lay the groundwork for defining the connection between these multivariate R^2_ms and the multivariate test statistics for Pillai's, Wilks', Hotelling's, and Roy's criteria (i.e., V, Λ, T, and θ). These four multivariate tests are all constructed in slightly different ways—depending on how the author chose to define the connection between R^2_m, its test statistic, and an F-test approximation to the test statistic. These four multivariate test statistics can be approached by the differing arithmetic involved in each, keeping in mind that the goal is to develop a bounded scalar measure of proportion of variance accounted for, $0 \le R^2_m \le 1$, which is connected to its own probability distribution.

To make the explanation more concrete, consider the univariate definition of $R^2_{hypothesis} = R^2_{full} - R^2_{restricted}$, resulting from a hypothesis imposed on a full model. In the univariate case, the value of $R^2_{hypothesis}$ is related to the $SS_{HYPOTHESIS}$ by $R^2_{hypothesis} * SS_{TOTAL}$, where the SS_{ERROR} is $(1 - R^2_{full}) * SS_{TOTAL}$. The F test has a close reciprocal relationship to these values and can easily be retrieved from $R^2_{hypothesis}$, $1 - R^2_{full}$, and degrees of freedom for hypothesis and error,

$$F_{(df_h, df_e)} = \frac{R^2_{hypothesis}}{1 - R^2_{full}} \cdot \frac{df_e}{df_h} = \frac{SS_{HYPOTHESIS}}{SS_{ERROR}} \cdot \frac{df_e}{df_h}, \qquad [4.24]$$

or conversely the value of $R^2_{hypothesis}$ can be retrieved from the value of F,

$$R^2_{hypothesis} = \frac{F(df_h)}{F(df_h) + df_e}. \qquad [4.25]$$

If the hypothesis SS of Equation 4.24 is equal to the full model SS, then the test is of the simultaneous full model R^2, whereas if the hypothesis SS is less than the full model SS the F test of Equation 4.24 is a test of the semipartial R^2. Conversely, a test of the partial R^2 is given by

$$F_{(df_h, df_e)} = \frac{R^2_{hypothesis}}{1 - R^2_{hypothesis}} \cdot \frac{df_e}{df_h}. \qquad [4.26]$$

These same relationships can be generalized to multivariate test statistics and their measures of strength of association. The summary of relationships between univariate and multivariate measures of association presented in Table 4.5 emphasize the fact that each of the multivariate test statistics and their value of R^2_m are generalizations of their univariate counterparts. In general, each multivariate test statistic consists of its own multivariate probability distribution, measure of association, and F-test approximation to the multivariate test statistic. The F-test approximations from all commercially available software default to a test of the *partial* R^2_m in any model, where $\mathbf{Q}_H \neq \mathbf{Q}_F$.[14] In the multivariate case, the general definition of the F-test approximation is

$$F_{(\nu_h, \nu_e)} = \frac{R^2_m}{1 - R^2_m} \cdot \frac{\nu_e}{\nu_h}, \qquad [4.27]$$

---

[14]If a test of the semipartial R^2_m is desired for a model in which $\mathbf{Q}_H \neq \mathbf{Q}_F$, the value of $R^2_{m(full)}$ for the full model is also required, and $1 - R^2_{m(full)}$ should replace the denominator of Equation 4.27.

Table 4.5 Comparison of Univariate and Multivariate Test Statistics and R^2_m

Test Statistic	*Multivariate Test Statistic and* R^2_m	*Univariate* R^2 *Conceptual Equivalent*	*Multivariate F-Test Equivalent*
Pillai's V	$V = Tr\left[\left(\mathbf{Q}_E + \mathbf{Q}_H\right)^{-1} \mathbf{Q}_H\right]$ $R^2_V = \frac{V}{s}$	$R^2_{partial}$ or R^2_{full} *if* $\mathbf{Q}_F = \mathbf{Q}_H$	$F_{(\nu_h, \nu_e)} = \frac{R^2_V}{1 - R^2_V} \cdot \frac{\nu_e}{\nu_h}$
Wilks' Λ	$\Lambda = \frac{\lvert\mathbf{Q}_E\rvert}{\lvert\mathbf{Q}_E + \mathbf{Q}_H\rvert}$ $R^2_\Lambda = 1 - \Lambda^{\frac{1}{s}}$	$1 - R^2_{partial}$ or $1 - R^2_{full}$ *if* $\mathbf{Q}_F = \mathbf{Q}_H$	$F_{(\nu_h, \nu_e)} = \frac{R^2_\Lambda}{1 - R^2_\Lambda} \cdot \frac{\nu_e}{\nu_h}$
Hotelling's T	$\mathrm{T} = Tr[\mathbf{Q}_E^{-1}\mathbf{Q}_H]$ $R^2_\mathrm{T} = \frac{\mathrm{T}}{\mathrm{T} + s}$	$\frac{R^2_{partial}}{1 - R^2_{partial}}$ $\frac{R^2_{full}}{1 - R^2_{full}}$ *if* $\mathbf{Q}_F = \mathbf{Q}_H$	$F_{(\nu_h, \nu_e)} = \frac{R^2_\mathrm{T}}{1 - R^2_\mathrm{T}} \cdot \frac{\nu_e}{\nu_h}$
Roy's θ	ρ^2_{max}	r^2	$F_{(\nu_h, \nu_e)} = \frac{\rho^2_{max}}{1 - \rho^2_{max}} \cdot \frac{\nu_e}{\nu_h}$

Note: ρ^2_{max} is the maximum squared canonical correlation coefficient between **Y** and **X**. The *F*-test approximations in the third column are the multivariate equivalents of the univariate *F* test and follows the *F* distribution with ν_h and ν_e degrees of freedom.

where the form of the approximation is the same as in the univariate case but with adjustments to the definitions of R_m^2 and degrees of freedom v_h and v_e that are specialized to each multivariate test statistic.

To develop tests of hypotheses based on Equation 4.27, it is first necessary to define R_m^2 for each of four multivariate test statistics. The differences in the definitions of these test statistics, summarized in Table 4.5, can be seen to depend on how the test's author uses R_m^2, or some function of R_m^2, to define a scalar test statistic related to their measure. Pillai's criterion uses R_m^2 directly, Wilks' criterion is defined in terms of $(1-R_m^2)$, and Hotelling's criterion depends on the quantity $\left(\frac{R_m^2}{1-R_m^2}\right)$. Roy's GCR does not depend on R_m^2 but is the largest squared canonical correlation between **Y** and **X**; we introduce Roy's criterion below and defer further explanation to Chapter 7. The fundamental differences between the R_m^2s and test statistics of Pillai, Wilks, Hotelling, and Roy result from differing ways of using the $\mathbf{Q}_H$ and $\mathbf{Q}_E$ matrices and one or more methods of reducing a matrix quantity to an interpretable scalar on the interval [0, 1]. We will refer to the conceptual and computational definitions displayed in Table 4.5 as we introduce the four multivariate test statistics and their measures of strength of association.

Pillai's Trace, *V*, and Its Measure of Association $\mathbf{R}_V^2$

For the linear model **Y** = **XB** + **E**, the multivariate test statistic based on Pillai's Trace criterion is a function of the error and hypothesis SSCP matrices. That is,

$$V = Tr[(\mathbf{Q}_E + \mathbf{Q}_H)^{-1}\,\mathbf{Q}_H], \qquad [4.28]$$

where the SSCP matrix $\mathbf{Q}_H$ is a function of the definition of a specific hypothesis leading to restrictions placed on the full model as described in the previous sections. Hence, $\mathbf{Q}_H$ can contain variability due to any hypothesized model ranging from the overall regression test of the full model to any partialled model as described in the discussion of Figure 4.2.

The quantity *V* of Equation 4.28 is a multivariate test statistic with a probability distribution based on the assumption of the multivariate normal distribution.[15] For the moment, we ignore the properties of the multivariate test statistics as test statistics, per se, and focus on their necessary role in the

[15]A discussion of the multivariate normal distribution and its role in evaluating multivariate test statistics is given in Rencher (1998, Chap. 2), and Tatsuoka (1988, Chap. 4).

definition of measures of association. In Chapter 5, we will introduce a version of the F-test approximation to V (and to Λ, T, and θ) that is widely used to test hypotheses represented by the matrix $\mathbf{Q}_H$ and that avoids the need to refer to specialized tables of the significance of the multivariate test statistics. Tables of V (Pillai, 1960) are available in Rencher (2002, table A11).

As seen in Table 4.5, every test statistic has a close connection to a measure of association that establishes the relationship between **Y** and **X**. The symmetric measure of association based on the Pillai's Trace (R_V^2) is very closely related to Hooper's $\bar{r}^2$ of Equation 4.21. Let q_h be the difference between the number of predictors in the full model (say, q_f) and the number of predictors in the restricted model imposed by a hypothesis (say, q_r). Furthermore, let s be the *smaller* of the two sets of p criterion or q_h predictor variables contained in the hypothesis SSCP matrix $\mathbf{Q}_H$,

$$s = \text{minimum}\left[p, q_h\right]. \qquad [4.29]$$

Hooper's Index is asymmetric, and if $p \neq q_h$ the index will yield different proportions of variance of **Y** accounted for by **X**, and **X** accounted for by **Y**. The measure of strength of association derived from Pillai's Trace V, R_V^2, is *symmetric* and is an *arithmetic* average based on the number of variables in the *smaller* of the two sets of variables documented by s,[16]

$$R_V^2 = \frac{Tr\left[(\mathbf{Q}_E + \mathbf{Q}_H)^{-1}\mathbf{Q}_H\right]}{s}. \qquad [4.30]$$

Note that R_V^2 is a scalar based on the conceptually equivalent ratio $\dfrac{\mathbf{Q}_H}{\mathbf{Q}_E + \mathbf{Q}_H}$ of Equation 4.18 and is identified as a *squared multivariate partial correlation* provided that the number of predictors in the full model, q_f, is greater than the number of predictor variables in the restricted model, q_r. As noted in earlier paragraphs, if $\mathbf{Q}_F = \mathbf{Q}_H$, then R_V^2 is not a partial correlation but is instead the full model proportion of variance in **Y** explained by **X**. In each of the multivariate test statistics reviewed here, the SSCP matrix $\mathbf{Q}_E$ is determined by the whole model fit of Equation 4.7.

Example 1: The Personality Data. Assessing the strength of the overall regression of the four dependent measures of the job application process on the three personality variables of the personality data is achieved by testing

---

[16]The value of s is based on the number of eigenvalues of $\left[(\mathbf{Q}_E + \mathbf{Q}_H)^{-1}\mathbf{Q}_H\right]$, which cannot be greater than the smaller of the variables in **Y** or the q_h predictors in **X**. The eigenvalue solution is introduced in Chapter 7.

the hypothesis that the 12 regression coefficients of the model, exclusive of the intercept, are drawn from a population in which the coefficients are zero,

$$\mathbf{H}_0: \begin{bmatrix} \beta_{01} & \beta_{02} & \beta_{03} & \beta_{04} \\ \beta_{11} & \beta_{12} & \beta_{13} & \beta_{14} \\ \beta_{21} & \beta_{22} & \beta_{23} & \beta_{24} \\ \beta_{31} & \beta_{32} & \beta_{23} & \beta_{34} \end{bmatrix} = \begin{bmatrix} \beta_{01} & \beta_{02} & \beta_{03} & \beta_{04} \\ 0 & 0 & 0 & 0 \\ 0 & 0 & 0 & 0 \\ 0 & 0 & 0 & 0 \end{bmatrix}.$$

Placing this restriction on the full model and taking the difference between the SSCP matrices for full and restricted models yields the hypothesis matrix $\mathbf{Q}_H$. Summing $\mathbf{Q}_H$ and $\mathbf{Q}_E$ from Table 4.1, computing $(\mathbf{Q}_E + \mathbf{Q}_H)^{-1}$, and evaluating Pillai's Trace by Equation 4.28 leads to the multivariate test statistic V,

$$V = .5062.$$

With $q_f = 3$ predictors in the full model and $q_r = 0$ predictors in the restricted model, the hypothesis model is based on $q_h = q_f - q_r = 3$ predictor variables, which implies that Pillai's Trace V should be averaged by $s =$ minimum[3, 4] to yield the value of R_V^2,[17]

$$R_V^2 = \frac{V}{s} = \frac{.5062}{3} = .169.$$

We conclude that about 17% of the joint variance of the dependent variables of background and social preparation, follow-up interview invitations, and job offers is accounted for by the joint variance of the personality dimensions of Neuroticism, Extraversion, and Conscientiousness. As a default first hypothesis for this multivariate analysis, we can say that there appears to be some visible connection between these employment outcomes and personality, but the exact nature of the relationship is yet unclear. It is often the case in complex models that the whole model association provides a beginning point for the analysis but does not provide information that is interpretively useful. Moreover, we do not yet know if this 17% of shared variance is a matter of statistical substance; if the null hypothesis were true, then the sample values of V and R_V^2 would be matters of sampling error; that is, the estimates of the parameters of the model would not be declared to be statistically significant by an appropriate test

[17]The value of Pillai's R_V^2 is equal to the smaller of Hooper's two indices of trace correlation.

statistic. In Chapter 5, we introduce a multivariate extension of the univariate general linear hypothesis test, similar to the univariate version in Equation 1.23, which embodies a convenient method for estimating $\mathbf{Q}_H$ and is suitable for testing myriad hypotheses about the multivariate relationships embedded in the whole model regression.

Example 2: The PCB Data. Evaluating the overall relationship of Example 2 between the six response variables (immediate and delayed visual memory, Stroop color and word tests, cholesterol and triglycerides) and the three predictor variables (age, gender, and PCB exposure) begins with the hypothesis that the 18 regression coefficients of the model whose parameters were estimated in Chapter 3 come from a population whose coefficients, exclusive of the intercept, are zero,

$$\mathbf{H}_0: \begin{bmatrix} \beta_{01} & \beta_{02} & \beta_{03} & \beta_{04} & \beta_{05} & \beta_{06} \\ \beta_{11} & \beta_{12} & \beta_{13} & \beta_{14} & \beta_{15} & \beta_{16} \\ \beta_{21} & \beta_{22} & \beta_{23} & \beta_{24} & \beta_{25} & \beta_{26} \\ \beta_{31} & \beta_{32} & \beta_{33} & \beta_{34} & \beta_{35} & \beta_{36} \end{bmatrix} = \begin{bmatrix} \beta_{01} & \beta_{02} & \beta_{03} & \beta_{04} & \beta_{05} & \beta_{06} \\ 0 & 0 & 0 & 0 & 0 & 0 \\ 0 & 0 & 0 & 0 & 0 & 0 \\ 0 & 0 & 0 & 0 & 0 & 0 \end{bmatrix}.$$

Placing this restriction on the full model and estimating the restricted model parameters leads to the restricted model SSCP matrix $\mathbf{Q}_R$, which is $\mathbf{0}$ for this hypothesis. Using $\mathbf{Q}_H = \mathbf{Q}_F$ and $\mathbf{Q}_E$ from Table 4.3, computing $(\mathbf{Q}_E + \mathbf{Q}_H)^{-1}$, and evaluating Equation 4.30, the value of Pillai's Trace for the PCB data is found to be

$$V = .4188$$

with multivariate R_V^2 equal to

$$R_V^2 = \frac{Tr\left[(\mathbf{Q}_E + \mathbf{Q}_H)^{-1}\mathbf{Q}_H\right]}{s} = .140.$$

Based on $q_h = q_f - q_r = 3$ predictors in the hypothesis model, the variables of age, gender, and PCB exposure account for about 14% of the joint variance in the criterion variables of memory (two variables), cognitive flexibility (two variables), and CVD risk factors (two variables). As was the case with the personality data, there is no especially interesting interpretation of this whole model regression, save for the fact that the proportion of variance accounted seems notable, although the statistical significance of its magnitude is yet to be ascertained. Moreover, since the dependent variables in this example are of two distinct classes (physical and psychological), more specialized hypotheses about the effect of the predictors will be worth investigating in the subsequent chapters.

Wilks' Λ and Its Measure of Association R^2_Λ

A second widely used multivariate test statistic, given by S. S. Wilks (1932), also approaches the solution to the multivariate problem by defining a test statistic (Λ) and its measure of strength of association that relies most heavily on the concept of unexplained variance captured in the $SSCP_{ERROR}$ matrix $\mathbf{Q}_E$. In the univariate case, a measure of R^2 can be approached indirectly from the proportion of variance in **Y** that is *not* explained by **X**. As seen in Equation 1.31, this value is given by

$$R^2_{Y \cdot X_1 X_2 \cdots X_q} = 1 - \frac{SS_{ERROR}}{SS_{TOTAL}}$$

if the full model is at issue. Squared partial correlations can be found similarly by

$$R^2_{YX_1|X_2 \cdots X_{q-1}} = 1 - \frac{SS_{ERROR}}{SS_{ERROR} + SS_{HYPOTHESIS}}.$$

Replacing the univariate SS in these definitions with their counterpart SSCP matrices $\mathbf{Q}_E$ and $\mathbf{Q}_H$ defines the conceptual equivalent of the multivariate measure of association of Wilks' Λ. To complete the generalization, we must define a test statistic (Λ) that relies on the coefficient of alienation, reduces the matrix quantities to scalar values that are faithful to the correlations among the several variables involved, and take an appropriate average across the s = minimum $[p, q_h]$ quantities involved in the hypothesis model to be fitted to the data.

Like all multivariate test statistics, the definition of the Λ involves matrices of order $(p \times p)$ as defined in the partition of Equation 4.7,

$$\Lambda = \frac{|\mathbf{Q}_E|}{|\mathbf{Q}_E + \mathbf{Q}_H|}. \qquad [4.31]$$

The determinant[18] of a matrix provides another method of reducing the several elements of these matrices down to scalars that faithfully capture in a single number the essence of the contents of the matrices and their ratio. Wilks (1932) defined a *generalized variance* as the determinant of a variance-covariance matrix. Since the SSCP matrices $\mathbf{Q}_E$ and $\mathbf{Q}_H$ are proportional

---

[18]The determinant of a matrix is a scalar that is uniquely associated with the elements of any square matrix. Methods for evaluating determinants are discussed in Fox (2009) and Schott (1997).

to variance-covariance matrices, their determinants are also proportional measures of generalized variance. Placed in these terms, Wilks' Λ is a ratio of the generalized variance of the errors of the multivariate linear model to the generalized variance of the "pseudo" total SSCP ($\mathbf{Q}_E + \mathbf{Q}_H$).[19] Hence $|\mathbf{Q}_E|$ captures the generalized variance of **Y** that is *unexplained* by the generalized variance in **X** and Λ is the *proportion* of the generalized variance in **Y** that is not accounted for by **X**. Consequently, Λ is a ratio of *unexplained-to-total generalized variance* and must therefore range between 0 and 1. When Λ = 1, none of the generalized variance in **Y** is accounted for by **X**. When Λ = 0, all the generalized variance in **Y** is accounted for by **X**. This suggests that a measure of strength of association with limits [0, 1] could be defined as $R^2_\Lambda = 1 - \Lambda$.

There is one additional complication in defining a measure such as $R^2_\Lambda = 1 - \Lambda$. The value of Λ is inherently based on a product series (determinants are computed as a function of successive products of elements), and a multivariate measure that does not overestimate the relationship should be based on an average of the product series. Such averages are achieved by the geometric mean, which applied to Λ is $\Lambda^{\frac{1}{s}}$. Consequently, a preferred measure of strength of association for Wilks' Λ that does not give an overestimate of the relationship is given by

$$R^2_\Lambda = 1 - \Lambda^{\frac{1}{s}}, \tag{4.32}$$

where s = minimum$[p, q_h]$ as was defined in the discussion of Pillai's *V*.

The multivariate sampling distribution of Λ has been tabled (Rencher, 2002, Table A9). It will be more convenient to refer Λ to an *F*-test approximation of Λ (Rao, 1951) that we will discuss in more detail in Chapter 5. We turn now to an illustration of Wilks' Λ for the personality data.

Example 1. For the personality data, we may use $\mathbf{Q}_H$ and $\mathbf{Q}_E$, from Table 4.1 and Equation 4.32 to find the value of Wilks' Λ for the whole model association

$$\Lambda = \frac{2.909 * 10^8}{5.121 * 10^8} = .5680$$

and for $s = 3$, the proportion of joint variance accounted for is

$$R^2_\Lambda = 1 - .5680^{\frac{1}{3}} = .172,$$

---

[19]If $\mathbf{Q}_F = \mathbf{Q}_H$, then the pseudo-total SSCP and total SSCP are identical.

which is slightly larger than the R_V^2 reported for Pillai's V.[20] While arithmetic and geometric means will not be equal to one another, the difference in substantive interpretation of 16.9% versus 17.2% of the joint variance in **Y** explained by the model is negligible. The two values would be equal when $s = 1$.

Example 2. For the PCB example data, we find from the relevant SSCP matrices from Table 4.3. Using the values of $\mathbf{Q}_E$ and $(\mathbf{Q}_E + \mathbf{Q}_H)$ we find

$$\Lambda = \frac{3.376*10^{17}}{5.583*10^{17}} = .6046$$

and the measure of strength of association for Λ based on $s = 3$ is

$$R_\Lambda^2 = 1 - .6046^{\frac{1}{3}} = .154.$$

Despite the minor differences noted above, this R_Λ^2 compares similarly to R_V^2 for Pillai's Trace—about 15% of the joint variance in **Y** is explained by the three predictors in model.

Hotelling's Trace, T, and Its Measure of Association, R_T^2

A third multivariate test statistic and its measure of association are due to Hotelling (1951). The multivariate test statistic is defined by the quantity

$$T = Tr\left[\mathbf{Q}_E^{-1}\mathbf{Q}_H\right], \quad [4.33]$$

which is the multivariate analogue of the ratio $SS_{HYPOTHESIS}/SS_{ERROR}$ that plays a key role in the definition of the univariate F test. The univariate quantity $SS_{HYPOTHESIS}/SS_{ERROR}$ is equal to $R^2/(1 - R^2)$, and it can be shown (Cramer & Nicewander, 1979; Haase, 1991) that the multivariate generalization of R^2 associated with Hotelling's Trace T is given by

$$R_T^2 = \frac{T}{T + s}, \quad [4.34]$$

where s is the minimum of p and q_h. The distribution of Hotelling's Trace is tabled by Rencher (2002, Table A12).

[20]One of the complications of multivariate analysis is that the four test statistics will not necessarily produce the same values of R_m^2 or F. In addition to different methods of averaging, other sources of the differences between multivariate test statistics are discussed in Olson (1974).

Example 1. Applying Equations 4.33 and 4.34 to the SSCP matrices of the personality data from Table 4.1 we find,

$$\mathrm{T} = .6357$$

and its measure of association is,

$$R_{\mathrm{T}}^2 = \frac{.6357}{.6357 + 3} = .175.$$

The substantive interpretation of these values follows the same pattern as for Pillai's and Wilks' criteria; about 18% of the joint variance in **Y** is explained by the model defined by the hypothesis (the full model in this case). The value of R_{T}^2 is of different magnitude than either R_V^2 or R_Λ^2. The difference is explained by the fact that Hotelling's Trace T in its underlying construction is based on a version of a weighted harmonic mean (Cramer & Nicewander, 1979; γ_3), which is inherently different from arithmetic and geometric means even when based on the same series of numbers. Despite these differences, the R_m^2 are reasonably similar for all three test statistics.

Example 2. The solution to Hotelling's T for the PCB data proceeds similarly. Evaluating Equation 4.33 on $\mathbf{Q}_E$ and $\mathbf{Q}_H$ from Table 4.3, we find

$$\mathrm{T} = .6156$$

and from Equation 4.34 the measure of strength of association for T is

$$R_{\mathrm{T}}^2 = \frac{.6156}{.6156 + 3} = .170.$$

The differing methods of averaging result in differing values of R_m^2 for these three multivariate test criteria, although the substantive conclusion is similar across all of them. Hotelling's Trace and R_{T}^2 reveal that about 17% of the joint, nonredundant variance in the entire set of response variables for this example is explained by age, gender, and exposure to PCBs. We pursue several interesting possibilities for evaluating and interpreting meaningful patterns of this overlap in the next chapter.

Roy's Greatest Characteristic Root θ, and Its Measure of Association, $r_{C_{max}}^2$

The fourth multivariate test statistic that is routinely reported by most software for multivariate analysis is Roy's (1957) GCR criterion, $\theta = r_{C_{max}}^2$. The closest univariate counterpart to $\theta = r^2_{C_{max}}$ is the square of the Pearson product-moment correlation coefficient. In Chapter 7, we introduce the solution to the eigenvalue problem, which is the basis of the technique of canonical correlation. The eigenvalues of Equation 4.21 that defined Hooper's trace index of correlation will be shown to be the squared canonical correlations between **Y** and **X**. For the present, we merely note that if we could form a weighted linear combination of the *Y* variables, say $l = \mathbf{aY} = a_1Y_1 + a_2Y_2 + \cdots + a_pY_p$, and a linear combination of the *X* variables, say $m = \mathbf{bX} = b_1X_1 + b_2X_2 + \cdots + b_qX_q$, and if we could choose the values of the vectors **a** and **b** to be "optimal" in the sense of rendering *l* and *m* maximally correlated, then the squared zero-order Pearson correlation, r^2_{lm}, is the maximum squared canonical correlation. It will be shown that the maximum eigenvalue of the quadruple product of Equation 4.20 is Roy's θ, which is the maximum squared canonical correlation, $r^2_{C_{max}}$. Tables of the significance of the maximum root criterion are given in Harris (2001, Table A.5) and an *F*-test approximation to $r^2_{C_{max}}$, which we introduce in Table 4.5, will receive greater attention in Chapter 5. Roy's $r^2_{C_{max}}$ GCR criterion is often substantially larger than R^2_V, R^2_Λ, or R^2_T, because it is an optimized empirical measure for which considerable shrinkage would be anticipated (Cohen & Nee, 1984). We will discuss these test statistics more extensively in Chapter 5 and canonical correlation in Chapter 7.

Examples 1 and 2: The Personality Data and the PCB Data. The maximum squared canonical correlation, that is, Roy's θ, for the personality data is $r^2_{C_{max}} = .246$, and for the PCB data is $r^2_{C_{max}} = .354$. In both cases, we observe that the proportion of joint variance in common between **Y** and **X** is noticeably larger than for the criteria of Pillai, Wilks, or Hotelling.[21] It would be prudent to be cautious about interpreting the nature of the relationship based only on the maximum root if it differs substantially in magnitude from any remaining characteristic roots for any given analysis. Olson (1976) discusses some of the conditions under which the four

---

[21]Because of this apparent inflation, the *F*-test approximation to Roy's θ is also very liberal and is said to be an upper bound. SPSS MANOVA refrains from printing the *F* test on θ, but SPSS GLM, SAS PROC REG, and STATA MANOVA all compute and print the value of *F* and *p* for Roy's GCR criterion. SAS 9.2 offers an option for a more conservative exact test of θ. It is also worth noting that SPSS MANOVA reports Roy's GCR as $r^2_{C_{max}}$, whereas SAS reports Roy's GCR as $\lambda = \frac{r^2_{C_{max}}}{1 - r^2_{C_{max}}}$. λ is the maximum eigenvalue of $\mathbf{Q}_E^{-1}\mathbf{Q}_H$. The relationship of λ *to* $r^2_{C_{max}}$ is dealt with more fully in Chapter 7.

multivariate measures of strength of association differ in concentrated and diffuse eigen structures.

Constructing Pillai's Trace V and Wilks' Λ From Univariate Regression Models

The matrix-based solutions to Pillai's Trace and Wilks' Λ given in Equations 4.28 to 4.32 are computationally efficient methods of solving for these test statistics and their measures of R_m^2. While R_V^2 and R_Λ^2 both behave as squared multiple correlations with a range of 0 to 1, there is an alternative way of constructing these two test statistics and their R_m^2s that can consolidate one's understanding of their meaning. The methods rely on successively residualizing the variables and computing univariate squared multiple correlations from these orthogonalized criterion and predictor variables. Since the residualizing is slightly different for Pillai's and Wilks' criteria, we deal with each separately.

Constructing Pillai's Trace and R_V^2 From Univariate R^2s

Pillai's Trace V can be shown to be the sum of p successively residualized response variables, each predicted from the full set of q explanatory variables in any multivariate model. Let $Y_2|Y_1$ be the residual $Y_2 - \hat{Y}_2$, removing the variance in Y_2 predicted from Y_1 from the regression $\hat{Y}_2 = \beta_0 + \beta_1 Y_1$. In a similar fashion, let $Y_3|Y_1Y_2$ be the residual of $Y_3 - \hat{Y}_3$ where $\hat{Y}_3$ is predicted from Y_1 and Y_2, and so on, finally letting $Y_p \mid Y_1 Y_2 \cdots Y_{p-1}$ be the residual of the last criterion variable, $Y_p - \hat{Y}_p$, where $\hat{Y}_p$ is predicted from the $p - 1$ preceding response variables. The predicted models and their residuals constructed from them are summarized in Table 4.6. These successively residualized response variables are orthogonal, which implies that for correlated variables the sum of squares of each residualized variable will be less than the sum of squares of its original version. Beginning with the variable Y_1, and following with the succeeding residuals $Y_2 \big| Y_1 \cdots Y_p \big| Y_1 Y_2 \cdots Y_{p-1}$, the p univariate squared multiple correlations with explanatory variables $X_1, X_2, \cdots, X_q$ can be evaluated as summarized in Table 4.6.

Unlike the redundancy coefficient, the sum of the values of the univariate R^2s defined in Column 4 of Table 4.6 involves no overlapping variance among the orthogonalized variables of **Y**. Consequently, the sum of the residualized univariate R^2s of Table 4.6 meets the conceptual definition of Pillai's Trace as the nonredundant joint variance in **Y** accounted for by **X**. This alternative computational form of V is therefore the sum of this series,

$$V = R^2_{Y_1 \cdot X_1 X_2 \cdots X_q} + R^2_{Y_2|Y_1 \cdot X_1 X_2 \cdots X_q} + \cdots + R^2_{Y_p|Y_1 \cdots Y_{p-1} \cdot X_1 X_2 \cdots X_q}. \quad [4.35]$$

Table 4.6 Variables, Residuals, and Univariate Full Model R^2s on the Successive Residuals for Pillai's V

Variable	*Residual*	*Predicted Model*	*Univariate* R^2
Y_1	—	—	$R^2_{Y_1 \cdot X_1X_2 \cdots X_q}$
$Y_2 \mid Y_1$	$Y_2 - \hat{Y}_2$	$\hat{Y}_2 = \beta_0 + \beta_1 Y_1$	$R^2_{Y_2\mid Y_1 \cdot X_1X_2 \cdots X_q}$
$Y_3 \mid Y_1Y_2$	$Y_3 - \hat{Y}_3$	$\hat{Y}_3 = \beta_0 + \beta_1 Y_1 + \beta_2 Y_2$	$R^2_{Y_3\mid Y_1Y_2 \cdot X_1X_2 \cdots X_q}$
⋮	⋮	⋮	⋮
$Y_p \mid Y_1Y_2 \cdots Y_{p-1}$	$Y_p - \hat{Y}_p$	$\hat{Y}_p = \beta_0 + \beta_1 Y_1 + \beta_2 Y_2 + \cdots + \beta_{p-1} Y_{p-1}$	$R^2_{Y_p\mid Y_1Y_2 \cdots Y_{p-1} \cdot X_1X_2 \cdots X_q}$

The value of V is symmetric whether based on partialled Y variables predicted from $\mathbf{X}$ or partialled X variables predicted from $\mathbf{Y}$. The average of V, however, is not symmetric for these two models. For reasons given earlier, the smaller of p or q_h is the appropriate divisor needed to arrive at a symmetric measure of multivariate strength of associaton. Hence,

$$R_V^2 = \frac{1}{s}V. \qquad [4.36]$$

Example 1. The correlation matrix for the residualized Y variables and $\mathbf{X}$ of the personality data are summarized in Table 4.7.[22]

The univariate R^2s of Equation 4.35 appear in the last column of Table 4.7. Their sum is

$$V = R^2_{Y_1 \cdot X_1X_2X_3} + R^2_{Y_2\mid Y_1 \cdot X_1X_2X_3} + \cdots + R^2_{Y_4\mid Y_1Y_2Y_3 \cdot X_1X_2X_3} = .5062.$$

Compare this value with the computed value of $V = Tr\left[\left(\mathbf{Q}_E^* + \mathbf{Q}_H^*\right)^{-1} \mathbf{Q}_H^*\right] = .5062$ from the worked example in previous paragraphs; the two results are equivalent. It follows then that the value of R_V^2 ($V/s = .169$) must be the same as that found by Equation 4.30. Note that

[22]The correlation matrix of the original variables for the Personality data and for the PCB data are given in Tables 2.1 and 2.2.

Table 4.7 Correlations Between Predictors and Residualized Response Variables for the Personality Data

	Y_1	$Y_2 \mid Y_1$	$Y_3 \mid Y_1Y_2$	$Y_4 \mid Y_1Y_2Y_3$	X_1	X_2	X_3	R^2
Y_1	1.000	.000	.000	.000	–.140	–.040	.270	.1002
$Y_2 \mid Y_1$	.000	1.000	.000	.000	–.034	.437	.117	.1920
$Y_3 \mid Y_1Y_2$	.000	.000	1.000	.000	–.013	.160	.312	.1038
$Y_4 \mid Y_1Y_2Y_3$	.000	.000	.000	1.0000	–.249	.153	–.075	.1102
X_1	–.140	–.034	–.013	–.249	1.000	–.100	–.200	
X_2	–.040	.437	.160	.153	–.100	1.000	.330	
X_3	.270	.117	.312	–.075	–.200	.330	1.000	

Note: Y_1 = background preparation, Y_2 = social preparation, Y_3 = follow-up interviews, Y_4 = offers, X_1 = Neuroticism, X_2 = Extraversion, X_3 = Conscientiousness.

this value is the second version of Hooper's $\bar{r}^2 = \frac{1}{q}Tr\left[\mathbf{R}_{YY}^{-1}\mathbf{R}_{YX}\mathbf{R}_{XX}^{-1}\mathbf{R}_{XY}\right]$ and since $q < p$ for these data, $\bar{r}^2 = R_V^2$ is based on an average of the smaller number of q or p variables in the model.[23] Pillai's Trace is therefore a sum of the proportions of variance in a set of orthogonal Y variables accounted for by $\mathbf{X}$. The interpretation of R_V^2 is therefore of an appropriate (i.e., s = minimum$[p,q]$) arithmetic average of those proportions. The relationship of this solution to the matrix solution of V and R_V^2 can be seen by noting that orthogonalizing the Y variables prior to analysis reduces the matrix $\mathbf{R}_{YY}$ in Equation 4.20 to an identity matrix. Setting $\mathbf{R}_{YY} = \mathbf{I}$ reduces Equation 4.20 to the matrix $\mathbf{Q}_F^*$ whose diagonal elements are the residualized R^2's as described above. The trace of $\mathbf{I}\mathbf{R}_{YX}\mathbf{R}_{XX}^{-1}\mathbf{R}_{XY}$ is the sum of those diagonal elements—Pillai's Trace V.

Example 2. The correlation matrix of the successively residualized Y variables and the explanatory variables $X_1, X_2,$ and X_3 for the PCB data are shown in Table 4.8. The correlation matrix of the original variables is presented in Table 2.2. The value of V for the PCB data is found by summing the $p = 6$ univariate R^2s of the last column of Table 4.8,

$$V = R^2_{Y_1 \cdot X_1 X_2 X_3} + R^2_{Y_2|Y_1 \cdot X_1 X_2 X_3} + \cdots + R^2_{Y_6|Y_1 Y_2 Y_3 Y_4 Y_5 \cdot X_1 X_2 X_3} = .4188,$$

and with $q < p = s = 3$, the value of $R_V^2 = .140$, which are the same values of V and R_V^2 found by the matrix methods of previous sections. Had we regressed $X_1 X_2 | X_1$, and $X_3 | X_1 X_2$ on the Y variables of this problem the value of V would be .4188 and Hooper's $\bar{r}^2$ would be equal to R_V^2.

The requirement that the divisor be defined as $s = \text{minimum}[p,q]$ is governed by the number of canonical correlations contained in the model $\mathbf{Y} = \mathbf{XB} + \mathbf{E}$; we delay the canonical correlation solution to the problem until Chapter 7.

Constructing Wilks' Λ and R_Λ^2 From p Univariate R_s^2

Reproducing Wilks' Λ from successively partialled variables follows much the same pattern as Pillai's V. Whereas the partialling process for V involved only the Y variables, the partialling process for Λ includes partialling of the Y variables from *both* the succeeding Y variables *and* the succeeding

[23]Residualizing the X variables, predicting the residuals from $\mathbf{Y}$, and summing the q univariate R^2s would give $V = .5062$ as V is symmetric and not affected by the direction of prediction. The asymmetry in $\bar{r}^2$ is corrected by averaging V across $s = \text{minimum}[p,q]$.

Table 4.8 Correlations Between Predictors and Residualized Response Variables for the PCB Data

	Y_1	$Y_2 \mid Y_1$	$Y_3 \mid Y_1Y_2$	$Y_4 \mid Y_1Y_2Y_3$	$Y_5 \mid Y_1 \cdots Y_4$	$Y_6 \mid Y_1 \cdots Y_5$	X_1	X_2	X_3	R^2
Y_1	1.00	0.00	0.00	0.00	0.00	0.00	–.387	–.043	–.364	.1668
$Y_2 \mid Y_1$	0.00	1.00	0.00	0.00	0.00	0.00	–.116	.106	–.142	.0293
$Y_3 \mid Y_1Y_2$	0.00	0.00	1.00	0.00	0.00	0.00	–.114	.149	–.086	.0382
$Y_4 \mid Y_1Y_2Y_3$	0.00	0.00	0.00	1.00	0.00	0.00	–.149	.042	–.104	.0251
$Y_5 \mid Y_1 \cdots Y_4$	0.00	0.00	0.00	0.00	1.00	0.00	.288	.021	.313	.1076
$Y_6 \mid Y_1 \cdots Y_5$	0.00	0.00	0.00	0.00	0.00	1.00	.147	–.123	.205	.0518
X_1	–.387	–.116	–.114	–.149	.288	.147	1.00	.047	.731	
X_2	–.043	106	.149	.042	.021	–.123	.047	1.00	–.130	
X_3	–.364	–.142	–.086	–.104	.314	.205	.731	–.130	1.00	

Note: Y_1 = immediate recall, Y_2 = delayed recall, Y_3 = Stroop color, Y_4 = Stroop word, Y_5 = log cholesterol, Y_6 = log triglycerides, X_1 = age, X_2 = gender, X_3 = PCBs.

X variables involved in each regression model. The variables $Y_1, Y_2|Y_1, \cdots, Y_p|Y_1 \cdots Y_{p-1}$ are the same as introduced above for Pillai's V. In addition, new sets of predictor variables must be constructed, each partialled from the successive Y variables—$(X_1|Y_1, X_2|Y_1, X_3|Y_1, \cdots, X_q|Y_1)$, $(X_1|Y_1Y_2, X_2|Y_1Y_2, X_3|Y_1Y_2, \cdots, X_q|Y_1Y_2)$, $\cdots$, $(X_1|Y_1 \cdots Y_{p-1}, X_2|Y_1 \cdots Y_{p-1}, \cdots, X_q | Y_1 \cdots Y_{p-1})$. The partialled variables, residuals, and prediction models are shown in Columns 1 to 3 of Table 4.9, and the univariate R^2s needed to construct Wilks' Λ are defined in Column 4. The definition of Wilks' $\Lambda = \frac{|\mathbf{Q}_E|}{|\mathbf{Q}_E + \mathbf{Q}_H|}$ as a ratio of error SSCP to the pseudo-total SSCP dictates that Λ is conceptually equivalent to a coefficient of alienation, $1 - R^2$. The further fact that the multivariate R^2_Λ is a geometric mean of a product series suggests that Λ could be approximated by a product series of the p successively partialled univariate R^2s defined in the last column of Table 4.9.

Thus, a symmetric measure of Λ is defined as a product series of the univariate $(1 - R^2)$s,

$$\Lambda = (1 - R^2_{Y_1 \cdot X_1 X_2 \cdots X_q})(1 - R^2_{(Y_2|Y_1) \cdot X_1|Y_1 X_2|Y_1 \cdots X_q|Y_1}) \cdots (1 - R^2_{(Y_p|Y_1 \cdots Y_{p-1}) \cdot X_1|Y_1 \cdots Y_{p-1} \cdots X_p|Y_1 \cdots Y_{p-1}}). \qquad [4.37]$$

The corresponding measure of R^2_Λ, based on a geometric mean to the power $\frac{1}{s}$ would be the equivalent of the value previously defined for Λ,

$$R^2_\Lambda = 1 - \Lambda^{\frac{1}{s}}. \qquad [4.38]$$

Example 1. The value of Λ and R^2_Λ can be reconstructed for the personality data from the residualized correlations of Table 4.10.

Applying the definition of Λ and R^2_Λ as a product series of k coefficients of alienation we have

$$\Lambda = \prod\left(1 - R^2_k\right) = (1 - .1002)(1 - .1928)(1 - .1139)(1 - .1174) = .5680 \cdot$$

Since the value of Λ is a product series, the proportion of joint variance in **Y** accounted for by **X** is estimated by the geometric mean based on $s = \min[p, q] = 3$. Hence, $R^2_\Lambda = 1 - \Lambda^{\frac{1}{s}} = .172$. The values of Λ and R^2_Λ computed from these fully partialled univariate regression models are identical to the results obtained by Equations 4.31 and 4.32.

Table 4.9 Variables, Residuals, and Univariate Full Model R^2s on the Residuals Needed for Wilks' Λ

Variable	*Residual*	*Predicted Model*	*Univariate* R^2
Y_1	—	—	$R^2_{Y_1 \cdot X_1 X_2 \cdots X_q}$
$Y_2 \mid Y_1$	$Y_2 - \hat{Y}_2$	$\hat{Y}_2 = \beta_0 + \beta_1 Y_1$	$R^2_{(Y_2 \mid Y_1) \cdot X_1 \mid Y_1 \cdots X_q \mid Y_1}$
$Y_3 \mid Y_1 Y_2$	$Y_3 - \hat{Y}_3$	$\hat{Y}_3 = \beta_0 + \beta_1 Y_1 + \beta_2 Y_2$	$R^2_{(Y_3 \mid Y_1 Y_2) \cdot X_1 \mid Y_1 Y_2 \cdots X_q \mid Y_1 Y_2}$
⋮	⋮	⋮	⋮
$Y_p \mid Y_1 Y_2 \cdots Y_{p-1}$	$Y_p - \hat{Y}_p$	$\hat{Y}_p = \beta_0 + \beta_1 Y_1 + \beta_2 Y_2 + \cdots + \beta_{p-1} Y_{p-1}$	$R^2_{(Y_p \mid Y_1 Y_2 \cdots Y_{p-1}) \cdot X_1 \mid Y_1 Y_2 \cdots Y_{p-1} \cdots X_q \mid Y_1 Y_2 \cdots Y_{p-1} \ldots X_q}$
$X_1 \mid Y_1$	$X_1 - \hat{X}_1$	$\hat{X}_1 = \beta_0 + \beta_1 Y_1$	
$X_2 \mid Y_1$	$X_2 - \hat{X}_2$	$\hat{X}_2 = \beta_0 + \beta_1 Y_1$	
⋮	⋮	⋮	⋮
$X_q \mid Y_1$	$X_q - \hat{X}_q$	$\hat{X}_q = \beta_0 + \beta_1 Y_1$	

Variable	*Residual*	*Predicted Model*	*Univariate* R^2
$X_1 \mid Y_1Y_2$	$X_1 - \widehat{X}_1$	$\widehat{X}_1 = \beta_0 + \beta_1 Y_1 + \beta_2 Y_2$	
$X_2 \mid Y_1Y_2$	$X_2 - \widehat{X}_2$	$\widehat{X}_2 = \beta_0 + \beta_1 Y_1 + \beta_2 Y_2$	
$\vdots$	$\vdots$	$\vdots$	
$X_q \mid Y_1Y_2$	$X_q - \widehat{X}_q$	$\widehat{X}_q = \beta_0 + \beta_1 Y_1 + \beta_2 Y_2$	
$\vdots$	$\vdots$	$\vdots$	
$X_1 \mid Y_1Y_2 \ldots Y_p$	$X_1 - \widehat{X}_1$	$\widehat{X}_1 = \beta_0 + \beta_1 Y_1 + \beta_2 Y_2 + \cdots + \beta_p Y_p$	
$X_2 \mid Y_1Y_2 \ldots Y_p$	$X_2 - \widehat{X}_2$	$\widehat{X}_2 = \beta_0 + \beta_1 Y_1 + \beta_2 Y_2 + \cdots + \beta_p Y_p$	
$\vdots$	$\vdots$	$\vdots$	
$X_q \mid Y_1Y_2 \ldots Y_p$	$X_q - \widehat{X}_q$	$\widehat{X}_q = \beta_0 + \beta_1 Y_1 + \beta_2 Y_2 + \cdots + \beta_p Y_p$	

Table 4.10 Correlations Needed to Construct Wilks' Λ for the Personality Data

	X_1	X_2	X_3	$X_1 \mid Y_1$	$X_2 \mid Y_1$	$X_3 \mid Y_1$	
Y_1	−.140	−.040	.270	.000	.000	.000	
$Y_2 \mid Y_1$	−.034	.437	.117	−.035	.438	.122	
$Y_3 \mid Y_1Y_2$	−.013	.160	.312	−.013	.160	.324	
$Y_4 \mid Y_1Y_2Y_3$	−.249	.153	−.075	−.251	.154	−.078	
	$X_1 \mid Y_1Y_2$	$X_2 \mid Y_1Y_2$	$X_3 \mid Y_1Y_2$	$X_1 \mid Y_1Y_2Y_3$	$X_2 \mid Y_1Y_2Y_3$	$X_3 \mid Y_1Y_2Y_3$	R^2
Y_1	.000	.000	.000	.000	.000	.000	.1002
$Y_2 \mid Y_1$	.000	.000	.000	.000	.000	.000	.1928
$Y_3 \mid Y_1Y_2$	−.013	.178	.326	.000	.000	.000	.1139
$Y_4 \mid Y_1Y_2Y_3$	−.251	.171	−.079	−.251	.174	−.083	.1174

Note: $\mathbf{R}_{YY} = \mathbf{I}$ and is omitted from the table. Y_1 = background preparation, Y_2 = social preparation, Y_3 = follow-up interviews, Y_4 = offers, X_1 = Neuroticism, X_2 = Extraversion, X_3 = Conscientiousness.

Example 2. The same pattern of fully partialled Y and X variables for the PCB data can be used to construct six univariate squared multiple correlations: $R^2_{immediate\ recall} = .1668$, $R^2_{delayed\ recall} = .0323$, $R^2_{Stroop\ word} = .0407$, $R^2_{Stroop\ color} = .0300$, $R^2_{cholesterol} = .1348$, and $R^2_{triglycerides} = .0686$. The product of the successively partialled coefficients of alienation from these R^2s is

$$\Lambda = (1-.1668)(1-.0323)(1-.04007)(1-.0300)(1-.1348)(1-.0686)$$
$$= .6046$$

and

$$R^2_\Lambda = 1-.6046^{\frac{1}{3}} = .154,$$

which agree with the values of the more computationally efficient ratio of matrix determinants.

These examples have been intended to give an alternative definition of Pillai's R^2_V and Wilks' R^2_Λ—they are literally an average (arithmetic or geometric) of a set of appropriately partialled univariate R^2s, and as such provide another perspective on how these multivariate measures are generalizations of their univariate counterparts. As suggested by Cramer and Nicewander (1979), we suspect that Hotelling's Trace T has a similar connection to the harmonic mean of a set of univariate R^2s—these differences in arithmetic partially explain why the R^2_m values of V, Λ, and T will often differ in practice.

We have shown that the partition of the SSCP matrices for a multivariate linear model provides the basis for obtaining multivariate test statistics as well as their measures of strength of association. With these estimates in hand, we can now evaluate whether the observed sample relationship between **Y** and **X** may be an artifact of sampling variability. In Chapter 5, we turn to a discussion of some general methods for testing a hypotheses about the relationship between **Y** and **X**. We introduce the multivariate general linear hypothesis test (Burdick, 1982; Rencher, 1998, pp. 272–273) as an efficient and flexible strategy for testing a diverse set of hypotheses on the multivariate linear model.

CHAPTER 5. TESTING HYPOTHESES IN THE MULTIVARIATE GENERAL LINEAR MODEL

The strategy for hypothesis testing in multivariate linear model analysis is based on the same four-step process described in Chapter 1 for univariate regression analysis—a model is specified, the parameters of the model are estimated, a measure of R_m^2 is computed, and hypotheses are tested by an appropriate test statistic. In the previous chapters, we have estimated the parameters of the multivariate model, partitioned the SSCP matrices of the model, and introduced four methods of defining a multivariate measure of association based on the test statistics of Pillai, Wilks, Hotelling, and Roy. The values of multivariate R_m^2s connected to these test statistics were shown to be functions of the SSCP matrices $\mathbf{Q}_T, \mathbf{Q}_F, \mathbf{Q}_E, \mathbf{Q}_R$, and $\mathbf{Q}_H$.

We noted in Chapter 4 that the definition of $\mathbf{Q}_H = \mathbf{Q}_F - \mathbf{Q}_R$, which is a key feature of all four multivariate test statistics and their R_m^2s, results from placing restrictions on $\mathbf{B}_{(q+1 \times p)}$ implied by a particular hypothesis, H_0. A test of significance of any R_m^2 defined by $\mathbf{Q}_H$ will allow the data analyst to eliminate sampling variation as an explanation for any nonzero sample value of $R_V^2, R_\Lambda^2, R_\mathrm{T}^2$, or R_θ^2.

In Chapter 1, we introduced and illustrated the general linear hypothesis test H_0:$\mathbf{L\beta} = \mathbf{k}$, in which a matrix $\mathbf{L}$ of contrasts can be written to identify a collection of one or more of the elements of $\boldsymbol{\beta}$ that are connected to a specific hypothesis to be tested. It was shown that the computation of the $SS_{\text{HYPOTHESIS}}$ by the use of $\mathbf{L}\hat{\boldsymbol{\beta}}$ (Equation 1.22) required in the numerator of an F test (Equation 1.23) for evaluating the hypothesis was the equivalent of the $SS_{\text{HYPOTHESIS}}$ extra-sums-of-squares approach used to formulate the difference between full and restricted model R^2s. This difference, $R^2_{hypothesis} = R^2_{full} - R^2_{restricted}$, was also shown to give the same numerator of the F test (Equation 1.18).

The use of a contrast matrix, $\mathbf{L}$, the parameter matrix, $\mathbf{B}$, and its sample estimate, $\hat{\mathbf{B}}$, can be used in precisely the same way to define the multivariate general linear test. This method of evaluating the hypothesis SSCP matrix, $\mathbf{Q}_H$, when referred to an F-test approximation to the four multivariate test statistics will prove to be a flexible procedure for testing a wide variety of multivariate hypotheses.

The Multivariate General Linear Test

The discussions of Chapter 4 established that partitioning the total SSCP into its constituent parts leads to the definition of four multivariate test statistics along with their respective measures of strength of association. We have shown that for any multivariate linear model with p response variables and q_f explanatory variables, the hypothesis SSCP matrix, $\mathbf{Q}_H$, based on q_h predictor variables, and the error SSCP matrix, $\mathbf{Q}_E$, are central to forming each of these multivariate test statistics and their measures of R^2_m. The definitions of the multivariate test statistics and their multivariate R^2_ms are summarized in Table 4.5.

Recall that $\mathbf{Q}_H$ is a function of the difference between a full model as defined by a specified design matrix **X**, and a restricted model $\mathbf{X}_R$, derived from imposing the constraints implied by H_0 on the full model. Since the specification of the design matrix **X** does not differ from univariate to multivariate analysis, the restrictions under H_0 on the rows of **B** are the same as those illustrated in Chapter 1 for the univariate case. The major difference in the definition of an appropriate contrast matrix for a multivariate model is that the coefficient matrix **B** has $p > 1$ columns and is of order $(q + 1 \times p)$ rather than $(q + 1 \times 1)$.

Using the $p = 2$, $q + 1 = 3$ example from Chapter 4, the hypothesis that the four regression coefficients of the two predictors (exclusive of the intercept) on the two dependent variables are from a null population is given by

$$H_0\!: \mathbf{B} = \begin{bmatrix} \beta_{01} & \beta_{02} \\ \beta_{11} & \beta_{12} \\ \beta_{21} & \beta_{22} \end{bmatrix} = \begin{bmatrix} \beta_{01} & \beta_{02} \\ 0 & 0 \\ 0 & 0 \end{bmatrix}.$$

Imposing these restrictions on the full model leads to the definition of the restricted model parameter matrix $\mathbf{B}_R = \begin{bmatrix} \beta_{01} & \beta_{02} \end{bmatrix}$. Substituting the sample estimates $\widehat{\mathbf{B}}_R$ for $\mathbf{B}_R$, we obtain the restricted model SSCP matrix $\mathbf{Q}_R = \widehat{\mathbf{B}}'_R \mathbf{X}'_R \mathbf{Y}$, and finally arrive at the required hypothesis SSCP, $\mathbf{Q}_H = \mathbf{Q}_F - \mathbf{Q}_R$. Once obtained, $\mathbf{Q}_H$ can be used to define any of the four multivariate test statistics and R^2_ms of Table 4.5.

A Convenient Method of Obtaining $\mathbf{Q}_H$. Defining a $(q_h \times q + 1)$ contrast matrix **L** and a contrast defined by the matrix product **LB** achieves the same result as the statement of H_0 above. That is,

$$H_0\!: \mathbf{LB} = \begin{bmatrix} 0 & 1 & 0 \\ 0 & 0 & 1 \end{bmatrix} \begin{bmatrix} \beta_{01} & \beta_{02} \\ \beta_{11} & \beta_{12} \\ \beta_{21} & \beta_{22} \end{bmatrix} = \begin{bmatrix} \beta_{11} & \beta_{12} \\ \beta_{21} & \beta_{22} \end{bmatrix} = \begin{bmatrix} 0 & 0 \\ 0 & 0 \end{bmatrix}.$$

Analogous to the univariate computation of the $SS_{HYPOTHESIS}$ of Equation 1.22, the multivariate hypothesis SSCP matrix $\mathbf{Q}_H$ can be similarly defined as a function of the contrast **LB**. Substituting $\hat{\mathbf{B}}$ for **B**, the $\mathbf{Q}_H$ consistent with the desired hypothesis is given by

$$\mathbf{Q}_H = \left(\mathbf{L}\hat{\mathbf{B}}\right)'\left(\mathbf{L}\left(\mathbf{X}'\mathbf{X}\right)^{-1}\mathbf{L}'\right)^{-1}\left(\mathbf{L}\hat{\mathbf{B}}\right). \qquad [5.1]$$

The matrix $\mathbf{Q}_H$ of Equation 5.1, and therefore any statistic that depends on $\mathbf{Q}_H$, is identical to the hypothesis SSCP matrix that would have been obtained by the equivalent, but more cumbersome, full-versus-restricted model, extra sums of squares approach to hypothesis testing.

An additional feature of the multivariate version of this contrast matrix that was not relevant to the univariate model entails possible contrasts on the p columns of **B** (i.e., the Y variables) in addition to the contrasts on the $q + 1$ rows of **B**. An extension to the multivariate general linear hypothesis test entails contrasts applied to *both* explanatory variables and the *response* variables. Such contrasts take the form

$$H_0: \mathbf{LBM} = \mathbf{K}. \qquad [5.2]$$

The matrix **M** can be introduced to accommodate a column-wise contrast among the p dependent variables to define an additional aspect of the test of hypotheses in the multivariate case. The matrix **M** is a key feature in the solution of repeated measures analysis of variance problems in which the Y variables are repetitions of the same measures across time or are multiple measures on the same scale for which contrasts are logically sensible, such as profile analysis (Maxwell & Delaney, 2004). If the variables Y_1 and Y_2 of a $p = 2$, $q + 1 = 3$ problem were repeated measures of the same scale assessed across time, then the contrast of the repeated measures factor (i.e., their mean difference $Y_1 - Y_2$) can be defined in **M**, and using the intercept X_0 as the sole predictor (i.e., not including the predictors X_1, X_2 in the model) can be achieved by testing the hypothesis defined by **LBM** as

$$H_0 : \mathbf{LBM} = [1]\begin{bmatrix}\beta_{01} & \beta_{02}\end{bmatrix}\begin{bmatrix}1 \\ -1\end{bmatrix} = \begin{bmatrix}\beta_{01} - \beta_{02}\end{bmatrix} = 0,$$

which is a test of the differences between the means of Y_1 and Y_2. More complex repeated measures problems for both between- and within-group factors can be handled by similar specifications of the contrasts in **L** and **M**. The computational algorithms in virtually all commercially available software will accommodate the use of the full **LBM** specification.

We point out these repeated measures applications to alert the reader to the extensive variety of models that can be analyzed by the general linear test strategy. The multivariate problems dealt with in this volume involve no repeated measures or profile contrasts on the columns of **B**, and we will ignore **M** altogether in the remainder of this volume and will rely solely on the specification of **LB** to test a variety of multivariate hypotheses. Comprehensive discussion of multivariate repeated measures analyses are given in O'Brien and Kaiser (1985) and Rencher (1998, pp. 296–297).

The Multivariate Test Statistics and Their F-Test Approximations

The multivariate sampling distributions for Pillai's Trace *V* (Pillai, 1960), Wilks' Λ (Wilks, 1932), Hotelling's Trace (Hotelling, 1951), and Roy's greatest characteristic root (GCR) θ (Roy, 1957) are very complicated and require special tables to cover the many possible combinations of p, q_h, *n,* and α that arise in practice.[1] As a practical matter, use of tables of *V,* Λ, T, and θ to perform tests of significance is cumbersome. There are, however, approximate tests of significance based on the familiar *F*-ratio that have been developed for hypothesis testing in the multivariate domain and that are sufficiently accurate for most applications. The best-known *F*-test approximation is given by Rao (1951), and this along with other similar *F*-test approximations have been incorporated into all commercial software for multivariate analysis. The multivariate *F*-test approximations to *V*, Λ, T, and θ are direct generalizations of the univariate *F*-test of Equations 4.24 to 4.26 and rely on a suitable measure of multivariate R_m^2,

$$F_{(\nu_h,\ \nu_e)} = \frac{R_m^2}{1-R_m^2} \cdot \frac{\nu_e}{\nu_h}. \qquad [5.3]$$

Replacing R_m^2 with any of the four multivariate measures of association of Table 4.5, and including multivariate degrees of freedom for error and hypothesis (ν_e and ν_h) appropriate to each test statistic provides the approximate *F*-test statistic of the hypothesis that generated the value of R_m^2, including tests of the full model and of partialled effects. This *F*-test approximation is very general and in some instances (if *p* or $q = 1$ or 2) produces

[1]Derivations are given in Anderson (2003, Sect. 8.5–8.6). Tables are available in Rencher (1998, Tables B.4-B.7) and extensive tables of Roy's θ are given in Harris (2001, Table A.5).

an exact test of the hypothesis. With the exception of Roy's GCR θ, the F-test approximations to V, Λ, and T are sufficiently accurate for most research applications.[2]

As noted in previous sections, the form of **LB** defines the hypothesis to be tested that can be the whole model test, partialled tests on single predictors, or partialled tests on more complex combinations of predictors. Any R_m^2 that depends on ($\mathbf{Q}_E + \mathbf{Q}_H$), rather than $\mathbf{Q}_T$, defines a partial, and not a semipartial, measure of multivariate association.[3] Most commercially available multivariate software procedures that provide a measure of R_m^2 as an output option will report the partialled result as a default. Other computational algorithms would be required to assess the multivariate semipartial R_m^2 (van den Burg & Lewis, 1988), which we do not pursue here.

In the paragraphs that follow, we will introduce the F-test approximations for V, Λ, T, and θ that can all be evaluated by the appropriate version of Equation 5.3. We will also attend to the definition of the multivariate degrees of freedom for hypothesis and error (denoted by ν_h, ν_e) for each test; the degrees of freedom are more involved than in the univariate case.

The *F*-Test Approximation for Pillai's Trace *V*

The F-test approximation for Pillai's Trace V (Olson, 1976; Pillai, 1960) is defined by substituting R_V^2 for R_m^2 in Equation 5.3:

$$F_{V(\nu_h,\nu_e)} = \frac{R_V^2}{1 - R_V^2} \cdot \frac{\nu_e}{\nu_h}. \qquad [5.4]$$

---

[2]The F-test approximation to θ is known to be a very liberal upper bound, and tables of θ will give more conservative results. Version 9.2 of SAS also offers a more conservative exact test of θ.

[3]Equation 5.3 and the R_m^2s of Table 4.5 define measures and tests on effects that are fully partialled, and not semipartialled, for any model other than the whole model analysis. If one desires the multivariate test on the semipartial R^2, such an effect would need to be programmed in SPSS MATRIX, SAS IML, STATA MATA, or similar matrix computer commands. Not all commercially available software will provide R_m^2 as part of their output; the value can be recovered by

$$R_m^2 = \frac{F(\nu_h)}{F(\nu_h) + \nu_e}.$$

The hypothesis and error degrees of freedom for the F-test approximation to Pillai's Trace depend on the values of n, p, q_f, q_h, and s:

$$\nu_h = pq_h,$$
$$\nu_e = s\left(n - q_f - 1 + s - p\right), \qquad [5.5]$$

where n is the sample size, q_f is the number of predictors in the full model (exclusive of the intercept), q_h is the number of rows in $\mathbf{L}$, p is the number of response variables, and s = minimum$[p, q_h]$.

Example 1. The null hypothesis that the $p = 4$ response variables of the personality data are unrelated to the full set of $q = 3$ predictor variables of Neuroticism, Extraversion, and Conscientiousness can be expressed as

$$H_0 : \mathbf{LB} = \begin{bmatrix} 0 & 1 & 0 & 0 \\ 0 & 0 & 1 & 0 \\ 0 & 0 & 0 & 1 \end{bmatrix} \begin{bmatrix} \beta_{01} & \beta_{02} & \beta_{03} & \beta_{04} \\ \beta_{11} & \beta_{12} & \beta_{13} & \beta_{14} \\ \beta_{21} & \beta_{22} & \beta_{23} & \beta_{24} \\ \beta_{31} & \beta_{32} & \beta_{33} & \beta_{34} \end{bmatrix}$$
$$= \begin{bmatrix} \beta_{11} & \beta_{12} & \beta_{13} & \beta_{14} \\ \beta_{21} & \beta_{22} & \beta_{23} & \beta_{24} \\ \beta_{31} & \beta_{32} & \beta_{33} & \beta_{34} \end{bmatrix} = \begin{bmatrix} 0 & 0 & 0 & 0 \\ 0 & 0 & 0 & 0 \\ 0 & 0 & 0 & 0 \end{bmatrix}.$$

Substituting $\hat{\mathbf{B}}$ for $\mathbf{B}$, evaluating $\mathbf{Q}_H$ by Equation 5.1, using $\mathbf{Q}_E$ from Table 4.3, and computing $V = Tr\left[\left(\mathbf{Q}_H + \mathbf{Q}_E\right)^{-1}\mathbf{Q}_H\right]$, we find $V = .5062$ and $R_V^2 = V/s = .169$. Noting that $\nu_e = 3(99 - 3 - 1 + 3 - 4) = 282$ and $\nu_h = 4(3) = 12$, we find the approximate F_V for Pillai's Trace V as

$$F_{V(12,\,282)} = \frac{.169}{1 - .169} \cdot \frac{282}{12} = 4.77,$$

with $p < .0001$.[4] We conclude from this result that the four job application response variables are significantly related to three personality characteristics. This full model relationship is not very detailed but is often reported as an

---

[4]The computed p value for an F of 4.77 on 12 and 282 degrees of freedom is $4.5*10^{-05}$. Noting that $p < .0001$ seems sufficiently informative.

initial evaluation of a multivariate relationship. In later sections, we introduce tests on substantively more interesting hypotheses about these data.

Example 2. The test of the whole model relationship between age, gender, and polychlorinated biphenyls (PCB) exposure and the set of six cognitive, memory, and cardiovascular disease (CVD) risk factor variables proceeds in a similar fashion. The null hypothesis of no relationship between the predictors and criteria of this example is specified as

$$H_0 : \mathbf{LB} = \begin{bmatrix} 0 & 1 & 0 & 0 \\ 0 & 0 & 1 & 0 \\ 0 & 0 & 0 & 1 \end{bmatrix} \begin{bmatrix} \beta_{01} & \beta_{02} & \beta_{03} & \beta_{04} & \beta_{05} & \beta_{06} \\ \beta_{11} & \beta_{12} & \beta_{13} & \beta_{14} & \beta_{15} & \beta_{16} \\ \beta_{21} & \beta_{22} & \beta_{23} & \beta_{24} & \beta_{25} & \beta_{26} \\ \beta_{31} & \beta_{32} & \beta_{33} & \beta_{34} & \beta_{35} & \beta_{36} \end{bmatrix}$$

$$= \begin{bmatrix} \beta_{11} & \beta_{12} & \beta_{13} & \beta_{14} & \beta_{15} & \beta_{16} \\ \beta_{21} & \beta_{22} & \beta_{23} & \beta_{24} & \beta_{25} & \beta_{26} \\ \beta_{31} & \beta_{32} & \beta_{33} & \beta_{34} & \beta_{35} & \beta_{36} \end{bmatrix} = \begin{bmatrix} 0 & 0 & 0 & 0 & 0 & 0 \\ 0 & 0 & 0 & 0 & 0 & 0 \\ 0 & 0 & 0 & 0 & 0 & 0 \end{bmatrix}.$$

The whole model hypothesis SSCP matrix $\mathbf{Q}_H$ is $\mathbf{Q}_F$ of Table 4.4. To test the null hypothesis on this $p = 6$, $q + 1 = 4$ example data set, we replace $\mathbf{B}$ with the estimates $\hat{\mathbf{B}}$ given in Chapter 3 and evaluate $\mathbf{Q}_H$ by Equation 5.1. Using the matrix $\mathbf{Q}_E$ from Table 4.4 and Equations 4.28 and 4.29, the value of Pillai's $V = .4189$ and $R_V^2 = .140$. The approximate F_V for the PCB data based on $n = 262$, $q_h = 3$, $\nu_e = 3(262 - 3 - 1 + 3 - 6) = 765$, and $\nu_h = 6(3) = 18$ is found to be

$$F_{V(18,\,765)} = \frac{.140}{1 - .140} \cdot \frac{765}{18} = 6.90,$$

with $p < .0001$. The 14% of the joint variance in this collection of response variables accounted for by age, gender, and PCB exposure is significantly different from zero. Hypotheses on the effect of PCB exposure, adjusted for age and gender, will be addressed in subsequent sections of this chapter.

The F-Test Approximation for Wilks' Λ

The evaluation of the statistical significance of Λ by Rao's (1951) F-test approximation follows a procedure similar to that for Pillai's Trace V. There is considerable similarity in the F-test approximations for all four multivariate test statistics: the hypotheses are the same, the contrasts $\mathbf{L}\hat{\mathbf{B}}$ are the same, and therefore the required SSCP matrices $\mathbf{Q}_H$ and $\mathbf{Q}_E$ are the same. Only the method of computing the test statistic and its R^2_m, as displayed in Table 4.5, differ across tests. The F-test approximation to Λ is a specialization of Equation 5.3,

$$F_{\Lambda(\nu_h,\nu_e)} = \frac{R^2_\Lambda}{1-R^2_\Lambda} \cdot \frac{\nu_e}{\nu_h}. \qquad [5.6]$$

The hypothesis degrees of freedom are identical to those for V, but the error degrees of freedom are more complicated and can yield noninteger values. The hypothesis and error degrees of freedom for F_Λ are

$$\nu_h = pq_h,$$

$$\nu_e = 1 + td - \frac{pq_h}{2}, \qquad [5.7]$$

where p and q_h have the usual meaning, and t and d are defined as

$$t = n - 1 - \frac{p + q_h + 1}{2},$$

$$d = \sqrt{\frac{p^2 q_h^2 - 4}{p^2 + q_h^2 - 5}}, \text{ if } p^2 q_h^2 \neq 4, \text{ otherwise } d = 1. \qquad [5.8]$$

Since the hypothesis to be tested by Wilks' Λ is the same as that tested by all four test statistics, the specification of $H_0 : \mathbf{LB} = 0$, and therefore the value of $\mathbf{Q}_H$, is already identified from previous paragraphs.

Example 1. For the personality, we find that $\Lambda = \dfrac{|\mathbf{Q}_E|}{|\mathbf{Q}_E + \mathbf{Q}_H|} = .5680$ with $R^2_\Lambda = 1 - \Lambda^{\frac{1}{3}} = .172$ for a whole model relationship based on $p = 4$ response variables and $q_h = 3$ predictors. From Equations 5.7 and 5.8, the value of $t = 94$, $d = 2.6458$, $\nu_h = 12$, and $\nu_e = 243.7$. The approximate F test for Wilks' Λ from Equation 5.6 is therefore

$$F_{\Lambda(12,243.7)} = \frac{.172}{1-.172} \cdot \frac{243.7}{12} = 4.84,$$

with $p < .0001$, and the same conclusion is reached as was drawn via Pillai's Trace V. Note that the values of R^2_m differ for the test statistics V and Λ; the two R^2_ms are based on an arithmetic and a geometric mean, respectively. Note also that ν_e is a noninteger value. While the differences between F-test approximations are not great for this example, it is frequently the case that the four test statistics will differ in absolute value. If the sample size is large relative to the value of R^2_m, then the powers of the tests will be sufficiently high and few contradictions should arise. Nonetheless, there will be times when the results from the four multivariate test statistics will lead to contradictory conclusions. Olson (1974, 1976) offers advice as to the relative robustness of the four test statistics and discusses the conditions that might lead to such differences.

Example 2. The F-test approximation to Λ for the PCB data can be evaluated from the results $\mathbf{Q}_H$ and $\mathbf{Q}_E$ from the example of Pillai's Trace. We find $\Lambda = \dfrac{|\mathbf{Q}_E|}{|\mathbf{Q}_E + \mathbf{Q}_H|} = .6046$ and $R^2_\Lambda = .154$ based on $p = 6$ response variables and $q_h = 3$ explanatory variables. From Equations 5.7 and 5.8, we find $t = 256$, $d = 2.8284$, $\nu_h = 18$, and $\nu_e = 716.08$. The F-test approximation is

$$F_{\Lambda(12,243.7)} = \frac{.154}{1-.154} \cdot \frac{716.08}{18} = 7.74,$$

with $p < .0001$, which leads to the same conclusion as reached by the F-approximation to Pillai's V. Wilks' Λ is among the oldest and the most widely cited multivariate test statistic in the research literature. Comparative tests of its power and robustness are reviewed in Olson (1974).

The F-Test Approximation for Hotelling's Trace

Hotelling's Trace is the third multivariate test statistic that is routinely calculated and reported by most commercial software for the analysis of multivariate data. The F-test approximation that is employed to evaluate the statistical significance of R^2_T has the same general structure as V and Λ,

$$F_{\mathrm{T}(\nu_h,\nu_e)} = \frac{R^2_\mathrm{T}}{1-R^2_\mathrm{T}} \cdot \frac{\nu_e}{\nu_h}, \qquad [5.9]$$

with degrees of freedom defined by

$$\nu_h = pq_h,$$

$$\nu_e = s(n - q_h - 2 - p) + 2. \qquad [5.10]$$

The solution to the F-test approximation to Hotelling's T follows a similar pattern to those for V and Λ. The solution is illustrated with the data from the two running examples.

Example 1. From $\mathbf{Q}_H$ and $\mathbf{Q}_E$, we find $\mathrm{T} = Tr\left[\mathbf{Q}_E^{-1}\mathbf{Q}_H\right] = .6357$ and $R_\mathrm{T}^2 = .175$ for the personality data based on $n = 99$, $p = 4$, $q_h = 3$, and $s = 3$. From Equations 5.9 and 5.10, $\nu_h = 12$ and $\nu_e = 272$ and the F-approximation to T is

$$F_{\mathrm{T}(12,272)} = \frac{.175}{1-.175} \cdot \frac{272}{12} = 4.80,$$

with $p < .0001$. The result is in close agreement with F_V and F_Λ.

Example 2. Analysis of the PCB example data from $\mathbf{Q}_H$ and $\mathbf{Q}_E$ of the previous examples is estimated from $\mathrm{T} = .6155$ and $R_\mathrm{T}^2 = .170$, based on $n = 262$, $p = 6$, $q_h = 3$, and $s = 3$. With $\nu_h = 18$ and $\nu_e = 755$, the F-test approximation to T is found as

$$F_{\mathrm{T}(18,755)} = \frac{.170}{1-.170} \cdot \frac{755}{18} = 8.61,$$

with $p < .0001$.

The results of Pillai's V, Wilks' Λ, and Hotelling's T all lead to the same conclusions about the whole model relationships embedded in both these examples; there is little to choose between the three tests. There is some evidence (Olson, 1974, 1976) that among these first three test statistics Pillai's V is somewhat less susceptible to violations of traditional assumptions underlying the test statistic, and is somewhat more powerful, but the differences, especially compared with Λ, appear to be minor in any practical sense. The fourth and final test statistic introduced in previous paragraphs that will be routinely found in computer output for multivariate analysis is Roy's GCR, θ.

F-Test Approximation to Roy's Greatest Characteristic Root, θ

Roy's θ, the multivariate test statistic based on the GCR of $(\mathbf{Q}_E + \mathbf{Q}_H)^{-1}\mathbf{Q}_H$,[5] is the maximum squared canonical correlation between **Y** and **X**. For

---

[5]Some computer programs report Roy's GCR as the largest eigenvalue, λ_{max}, of $\mathbf{Q}_E^{-1}\mathbf{Q}_H$. The relationship between λ_{max} and $r_{C_{max}}^2$ is

$$r_{C_{max}}^2 = \frac{\lambda_{max}}{1+\lambda_{max}}$$

and is discussed at greater length in Chapter 7.

the sample value, $r^2_{C_{max}}$ the F-test approximation to θ is a special case of Equation 5.3,

$$F_{\theta_{(\nu_h, \nu_e)}} = \frac{r^2_{C_{max}}}{1 - r^2_{C_{max}}} \cdot \frac{\nu_e}{\nu_h} \qquad [5.11]$$

with degrees of freedom defined as

$$\nu_h = p,$$
$$\nu_e = n - q_f - 1 - p + q_h. \qquad [5.12]$$

Although we delay explanation of the computational details of θ and $r^2_{C_{max}}$ to Chapter 7, the F-test approximation is introduced and illustrated here because it is printed by most statistical software for multivariate analysis.

Example 1. For the personality data, the largest squared canonical correlation is found to be $r^2_{C_{max}} = .2464$ based on $n = 99$, $p = 4$, and $q_h = 3$. From Equations 5.11 and 5.12, $\nu_h = 4$, $\nu_e = 94$, and the F-test approximation is

$$F_{\theta_{(4,94)}} = \frac{.2464}{1 - .2464} \cdot \frac{94}{4} = 7.69,$$

with $p < .0001$.

Example 2. Roy's θ for the PCB example is computed as $r^2_{C_{max}} = .3538$, from the model based on $n = 262$, $p = 6$, and $q_h = 3$. For these data, $\nu_h = 6$, $\nu_e = 255$, and the F-test approximation is

$$F_{\theta_{(6,255)}} = \frac{.3538}{1 - .3538} \cdot \frac{255}{6} = 23.27,$$

and $p < .0001$.

The F-test approximation to Roy's GCR is an upper-bound statistic and is known to be extremely liberal when compared with the other three test statistics in common usage. Olson (1974, 1976) has shown that the Type I error rates for Roy's θ is substantially inflated relative to those of V, Λ, and T. Two alternatives that lead to a more accurate test of hypotheses based on Roy's GCR include (1) the MSTAT=EXACT option in the SAS GLM procedure and (2) evaluation by tabled values of θ (e.g., Harris, 2001, Table A.5). For these examples, both procedures lead to the same rejections of the null hypothesis of the full model as presented in previous paragraphs. In models where the effects are of smaller magnitude, the exact tests (SAS or tables) would be preferred for evaluating Roy's θ.

The General Linear Test on Individual Predictors and Sets of Predictors

The whole model test of hypothesis presented in the previous sections is typically the first step in evaluating multivariate relationships between **Y** and **X**, even though the whole model may be the least interesting of many hypotheses that could be tested. Extending the strategy for testing hypothesis by the general linear test to partialled single predictors, or to partialled sets of predictors, greatly expands the number and type of hypotheses that can be tested. These specialized tests are supported by virtually all commercial software. We focus first on the multivariate tests of individual partialled effects, followed by procedures for testing hypotheses on sets of predictors and other complex hypotheses. In each section, we also introduce the univariate follow-up tests as one method of understanding the contribution of the response variables to the multivariate relationship.

Testing Multivariate Hypotheses on Individual Predictor Variables: The Personality Data

Consider the running example of the personality data in which four variables related to job interviews are to be predicted from three personality variables: Neuroticism, Extraversion, and Conscientiousness. Given the correlations of Table 2.1, it is known that these sets of characteristics covary; their average correlation in these data is approximately ±.30. From the whole model test, there is a significant relationship between the collection of predictors and the response variables, but it is unclear how each of the dimensions might be contributing to that overall relationship. Examining the relationship between each personality dimension without adjusting for the remaining personality factors would overestimate its unique contribution to **Y**. Testing the multivariate hypothesis of the effect of each predictor on **Y** while adjusting for the overlap (i.e., confounding) of the remaining predictors addresses this problem.

Neuroticism is a dimension of personality that embodies the characteristics of anxiety, worry, fear, self-consciousness, and vulnerability. It is not unreasonable to predict that persons who are higher on this dimension would have greater difficulty with all aspects of the interviewing process. Assuming that the predictors are X_0 = unit vector, X_1 = Neuroticism, X_2 = Extraversion, and X_3 = Conscientiousness, then the hypothesis on the unique effect of Neuroticism can be specified by appropriate contrast vector $\mathbf{L}$ on the parameters in $\mathbf{B}_{(4\times4)}$,

$$H_0: \mathbf{LB} = \begin{bmatrix} 0 & 1 & 0 & 0 \end{bmatrix} \begin{bmatrix} \beta_{01} & \beta_{02} & \beta_{03} & \beta_{04} \\ \beta_{11} & \beta_{12} & \beta_{13} & \beta_{14} \\ \beta_{21} & \beta_{22} & \beta_{23} & \beta_{24} \\ \beta_{31} & \beta_{32} & \beta_{33} & \beta_{34} \end{bmatrix}$$
$$= \begin{bmatrix} \beta_{11} & \beta_{12} & \beta_{13} & \beta_{14} \end{bmatrix} = \begin{bmatrix} 0 & 0 & 0 & 0 \end{bmatrix}.$$

Substituting the estimates of the parameters $\hat{\mathbf{B}}$ of Table 3.2 into Equation 5.1 and evaluating $\mathbf{Q}_H$ for Neuroticism adjusted for Extraversion and Conscientiousness we find

$$\mathbf{Q}_{H_{(N|E,C)}} = \begin{bmatrix} 14.345 & 6.536 & -.543 & 2.515 \\ 6.536 & 2.978 & -.247 & 1.146 \\ -.543 & -.247 & .021 & -.095 \\ 2.515 & 1.146 & -.095 & .441 \end{bmatrix},$$

with $q_h = 1$ since $\mathbf{L}$ contains a single row.

The matrix $\mathbf{Q}_E$ for the personality data is from Table 4.1. The sum $\mathbf{Q}_E + \mathbf{Q}_H$ provides the additional quantity necessary to compute all four multivariate test statistics from their definitions of Table 4.5. The multivariate test statistics, their measures of R^2_m, and their approximate F tests for the hypothesis that Neuroticism uniquely accounts for significant joint variability of the four response variables are presented in Table 5.1.

The hypotheses tests on the unique contribution of Extraversion adjusted for Neuroticism and Conscienc、iousness, and for Conscientiousness adjusted for Neuroticism and Extraversion are approached by the same process in which the contrast matrix $\mathbf{L}$ is chosen to state the hypothesis, $H_0: \mathbf{LB} = 0$, and the selection of parameter estimates by $\mathbf{L\hat{B}}$ leads to the appropriate hypothesis SSCP matrix. For the unique effect of extraversion, we write the hypothesis as

$$H_0: \mathbf{LB} = \begin{bmatrix} 0 & 0 & 1 & 0 \end{bmatrix} \begin{bmatrix} \beta_{01} & \beta_{02} & \beta_{03} & \beta_{04} \\ \beta_{11} & \beta_{12} & \beta_{13} & \beta_{14} \\ \beta_{21} & \beta_{22} & \beta_{23} & \beta_{24} \\ \beta_{31} & \beta_{32} & \beta_{33} & \beta_{34} \end{bmatrix}$$
$$= \begin{bmatrix} \beta_{21} & \beta_{22} & \beta_{23} & \beta_{24} \end{bmatrix} = \begin{bmatrix} 0 & 0 & 0 & 0 \end{bmatrix}$$

Table 5.1 Multivariate Tests of Partialled Individual Predictor Variables for the Personality Data

Hypothesis	*Test*	*Statistic*	R^2_m	ν_h, ν_e	*F-Approx.*	*p*
N | E, C	Pillai's *V*	.082	.082	4, 92	2.05	.094
	Wilks' Λ	.918	.082	4, 92	2.05	.094
	Hotelling's T	.089	.082	4, 92	2.05	.094
	Roy's θ	.082	.082	4, 92	2.05	.094
E | N, C	Pillai's *V*	.236	.236	4, 92	7.09	<.0001
	Wilks' Λ	.764	.236	4, 92	7.09	<.0001
	Hotelling's T	.308	.236	4, 92	7.09	<.0001
	Roy's θ	.308	.236	4, 92	7.09	<.0001
C | N, E	Pillai's *V*	.188	.188	4, 92	5.33	.001
	Wilks' Λ	.812	.188	4, 92	5.33	.001
	Hotelling's T	.232	.188	4, 92	5.33	.001
	Roy's θ	.188	.188	4, 92	5.33	.001

Note: All *F* tests are exact and equivalent for $s = 1$. *N* = Neuroticism, *E* = Extraversion, *C* = Conscientiousness.

and for the unique effect for conscientiousness we write

$$H_0\text{:}\mathbf{LB} = \begin{bmatrix}0 & 0 & 0 & 1\end{bmatrix}\begin{bmatrix}\beta_{01} & \beta_{02} & \beta_{03} & \beta_{04}\\ \beta_{11} & \beta_{12} & \beta_{13} & \beta_{14}\\ \beta_{21} & \beta_{22} & \beta_{23} & \beta_{24}\\ \beta_{31} & \beta_{32} & \beta_{33} & \beta_{34}\end{bmatrix}$$

$$= \begin{bmatrix}\beta_{31} & \beta_{32} & \beta_{33} & \beta_{34}\end{bmatrix} = \begin{bmatrix}0 & 0 & 0 & 0\end{bmatrix}.$$

Substituting the estimates of the parameters from Table 3.2 to define $\mathbf{L\hat{B}}$ for each hypothesis and evaluating Equation 5.1 leads to the two hypothesis SSCP matrices for Extraversion and Conscientiousness, respectively,

$$\mathbf{Q}_{H(E|N,C)} = \begin{bmatrix}32.643 & -91.298 & -3.926 & -6.640\\ -91.298 & 255.351 & 10.981 & 18.571\\ -3.926 & 10.981 & .472 & .799\\ -6.640 & 18.571 & .799 & 1.351\end{bmatrix}$$

and

$$\mathbf{Q}_{H(C|N,E)} = \begin{bmatrix}129.223 & 51.779 & 15.647 & -3.896\\ 51.779 & 20.748 & 6.270 & -1.561\\ 15.647 & 6.270 & 1.895 & -.472\\ -3.896 & -1.561 & -.472 & .117\end{bmatrix}.$$

Once $\mathbf{Q}_H$ is in hand, and using $\mathbf{Q}_E$ from the partition of the whole model, the multivariate test statistics, measures of R_m^2, and F-test approximations are readily obtained. The results are summarized in Table 5.1.

If we adopt a conventional level of $\alpha = .05$ for concluding any null hypothesis should be rejected, we see from the F-test approximations of Table 5.1 that Neuroticism, after adjusting for its collinearity with Extraversion and Conscientiousness is not declared statistically significant. There is insufficient

[6]For interpreting magnitude of effect, we single out Pillai's R_V^2 as a representative measure. With $q_h = 1$, all tests produce the same R_m^2. Substantial differences between different realizations of R_m^2 could, however, be observed.

evidence to conclude the vector of partial regression coefficients, and the partialled measures of R_m^2, came from a population whose values are nonzero. Conversely, the effects of Extraversion and Conscientiousness are seen to be statistically significant. The strength of association R_V^2 for extraversion ($R_V^2 = .24$) and conscientiousness ($R_V^2 = .19$)[6] suggest that a substantial amount of the joint variability of the interview behaviors are separately explained by each of these predictors. The nature of the relationship, while statistically significant at the level of the multivariate test, is still not entirely clear. While the contribution of the predictors to the overall relationship is resolved, we do not yet know how the (possibly) correlated dependent variables contribute to the significant multivariate effects. It is a common practice suggested by most authors to fit and interpret the p univariate regression models as an aid to understanding any statistically significant multivariate relationship. There is some evidence that interpreting the univariate follow-up tests *if, and only if*, the initial multivariate test is significant provides adequate protection for escalating experiment-wise error rates resulting from unadjusted, repetitive univariate tests (Rencher & Scott, 1990).[7]

Follow-Up Univariate R^2s and F-Tests for the Personality Data. The univariate regression models for each of the dependent variables, partialled in the same manner as dictated by the multivariate hypothesis on the predictors (i.e., defined by **L**), are easily obtained as part of the intermediate solution of the multivariate analysis. We noted in Chapter 4 that the diagonal elements of the partitioned SSCP matrices (always of order $p \times p$) carry all the information necessary to compute the univariate R^2s and F tests for each of the p dependent variables. The univariate follow-up tests for the three predictors of the personality data example are summarized in Table 5.2.

We would not interpret the univariate F-tests for any nonsignificant multivariate effect; the significant ($p < .05$) univariate F-test for Offers as a function of Neuroticism may be spuriously due to inflated Type I errors.[8] The partialled univariate F-tests for Extraversion and Conscientiousness, however, are associated with significant multivariate effects and suggest some interesting ways in which the dependent variables may contribute to the multivariate relationship.

---

[7]In Chapter 7, we discuss alternative methods of assessing the contribution of the dependent variables to the multivariate relationship that are sensitive to the dependent variable correlations.

[8]Controlling the Type I error rate by the Bonferroni split (i.e., $\alpha = .05/4 = .0125$) would also suggest a failure to reject the null hypothesis for this test.

Table 5.2 Univariate Follow-Up Tests for Neuroticism, Extraversion and Conscientiousness

Hypothesis	*Response Variable*	R^2	υ_h, υ_e	*F*	*p*
N | E, C	Background prep.	.009	1, 95	0.911	.342
	Social prep.	.001	1, 95	0.138	.711
	Follow-up interview	.001	1, 95	0.118	.731
	Offers	.042	1, 95	4.138	.045
E | N, C	Background prep.	.021	1, 95	2.072	.153
	Social prep.	.111	1, 95	11.821	.001
	Follow-up interview	.028	1, 95	2.720	.102
	Offers	.118	1, 95	12.674	.001
C | N, E	Background prep.	.079	1, 95	8.202	.005
	Social prep.	.010	1, 95	0.960	.330
	Follow-up interview	.103	1, 95	10.912	.001
	Offers	.011	1, 95	1.102	.296

Note: N = Neuroticism, E = Extraversion, C = Conscientiousness.

The personality dimension of Extraversion is characterized by warmth, gregariousness, activity, and positive emotions. The univariate *F*-tests of Table 5.2 suggest that Extraversion is primarily predictive of those job application behaviors that are social in nature: (1) social preparation in which the student sought information about potential employers by interpersonal acts—speaking with faculty, relatives, and friends or by contacting someone in company. Interestingly, background preparation that depends on assimilating information about a potential employer by noninteractive, impersonal sources (e.g., reading, annual reports, balance sheets) was unrelated to Extraversion and (2) the number of job offers received is significantly a function of Extraversion—extraverted personality types apparently have engaging interview styles of interpersonal interaction. The number of follow-up interviews was also not significantly related to Extraversion.

Conversely, the personality dimension of Conscientiousness (adjusted for Extraversion and Neuroticism) is seen to be significantly related to background preparation and the number of follow-up interviews received. Since Conscientiousness is characterized by competence, order, dutifulness, self-discipline, and achievement-striving, it is intuitively appealing to see that this factor predicts the factual preparation for the job search process and is unrelated to the more social aspects already predicted from Extraversion.

It should be kept in mind that the univariate follow-up tests are tests in which the dependent variables are *not* adjusted for their relationships with one another. Consequently, each univariate test may overestimate the contribution of a single response variable to the multivariate relationship.[9] In later chapters, we will introduce other methods of assessing predictor and response variable contributions to the multivariate relationship that adjust for the redundant variance in both **Y** and **X**.

Testing Multivariate Hypotheses on Individual Predictor Variables: The PCB Data.

The multivariate tests of individual predictor variables proceeds in the same fashion as illustrated for the personality data. The hypothesis is specified by the contrast on the parameters, $\mathbf{L}\hat{\mathbf{B}}$, from which the hypothesis SSCP matrix $\mathbf{Q}_H$ is obtained by substituting $\hat{\mathbf{B}}$ into Equation 5.1, and computing the multivariate statistics of Table 4.5. The contrast that defines the null hypothesis for PCB exposure, adjusted for age and gender, would be

[9]Roy-Bargman step-down *F* tests (Stevens, 2007, Chap. 10) have been proposed as one solution to the problem of correlated dependent variables.

$\mathbf{L} = [0 \quad 0 \quad 0 \quad 1]$. Postmultiplying the estimates of the parameters of Table 3.8 by $\mathbf{L}$ and evaluating Equation 5.1 leads to the required $\mathbf{Q}_H$,

$$\mathbf{Q}_{H_{(PCBs|age,\,gender)}} = \begin{bmatrix} 47.203 & 54.851 & -6.822 & -22.658 & -1.680 & -5.347 \\ 54.851 & 63.738 & -7.927 & -26.329 & -1.953 & -6.213 \\ -6.822 & -7.927 & .986 & 3.274 & .243 & .773 \\ -22.658 & -26.329 & 3.274 & 10.876 & .807 & 2.566 \\ -1.680 & -1.953 & .243 & .807 & .060 & .190 \\ -5.347 & -6.213 & .773 & 2.566 & .190 & .606 \end{bmatrix}.$$

The individual hypotheses for age ($\mathbf{L} = [0 \quad 1 \quad 0 \quad 0]$) and gender ($\mathbf{L} = [0 \quad 0 \quad 1 \quad 0]$), each adjusted for the remaining predictors can be similarly evaluated and yield hypothesis SSCP matrices of $\mathbf{Q}_{H_{(age|gender,\,PCBs)}}$ and $\mathbf{Q}_{H_{(gender|age,\,PCBs)}}$ (not shown here). The matrix $\mathbf{Q}_E$ is the same SSCP matrix used in tests of the full model of previous paragraphs. The multivariate results presented in Table 5.3 reveal that all three partialled effects are statistically significant at $\alpha = .05$.

The hypothesis on the PCB effect is the most important in this research as age and gender are known by previous research to be competing explanations for any disease process.

Follow-Up Univariate R^2s and F Tests for the PCB Data. The univariate follow-up tests, shown in Table 5.4, offer some clues as to the relative contribution of the six response variables to these multivariate relationships.

As is common in all health studies that involve exposure across time, age is an important covariate and possible confounder—this is clear in these data as the age effect influences all but one of the six response variables. The gender relationship is documented largely by the cognitive flexibility variables with little contribution emerging from the memory or CVD risk factors. The exposure to PCBs, which is the central thrust of this example, showed a significant multivariate effect after adjusting for both age and gender. The univariate follow-up tests of this partialled PCB effect suggests that this multivariate relationship is largely defined by the effects of PCBs on the memory and CVD risk factor response variables. The cognitive flexibility variables seem to make little contribution to understanding the multivariate relationship. Although these univariate F-tests do not take account of the correlations among the response variables, we will see in Chapter 7 that this interpretation is consistent with other methods that adjust for the overlap among dependent variables, namely, canonical correlation.

Table 5.3 Multivariate Tests of Partialled Individual Predictor Variables for the PCB Data

Hypothesis	*Test*	*Statistic*	R^2_m	υ_h, υ_e	*F-Approx.*	*p*
PCBs | A, G	Pillai's *V*	.078	.078	6, 253	3.58	.002
	Wilks' Λ	.922	.078	6, 253	3.58	.002
	Hotelling's T	.084	.078	6, 253	3.58	.002
	Roy's *θ*	.078	.078	6, 253	3.58	.002
A | G, PCBs	Pillai's *V*	.076	.076	6, 253	3.48	.003
	Wilks' Λ	.924	.076	6, 253	3.48	.003
	Hotelling's T	.083	.076	6, 253	3.48	.003
	Roy's *θ*	.076	.076	6, 253	3.48	.003
G | A, PCBs	Pillai's *V*	.052	.052	6, 253	2.31	.035
	Wilks' Λ	.948	.052	6, 253	2.31	.035
	Hotelling's T	.055	.052	6, 253	2.31	.035
	Roy's *θ*	.052	.052	6, 253	2.31	.035

Note: All *F*-tests are exact and equivalent for $s = 1$. PCBs = log transformed PCB exposure, A = age, *G* = gender.

Table 5.4 Univariate Follow-Up Tests for Age, Gender, and PCB Exposure

Hypothesis	*Response Variable*	R^2	ν_h, ν_e	F	p
PCBs | A, G	Visual memory–imm	.020	1, 258	5.17	.024
	Visual memory–del	.022	1, 258	5.80	.017
	Stroop word	.00001	1, 258	0.002	.966
	Stroop color	.0001	1, 258	0.03	.862
	Cholesterol	.033	1, 258	8.85	.003
	Triglycerides	.042	1, 258	11.43	.001
A | PCBs, G	Visual memory–imm	.030	1, 258	8.04	.005
	Visual memory–del	.026	1, 258	6.83	.009
	Stroop word	.020	1, 258	5.38	.021
	Stroop color	.036	1, 258	9.77	.002
	Cholesterol	.015	1, 258	3.96	.048
	Triglycerides	.007	1, 258	1.89	.171

(Continued)

Table 5.4 (Continued)

Hypothesis	*Response Variable*	R^2	ν_h, ν_e	*F*	*p*
G | A, PCBs	Visual memory–imm	.004	1, 258	0.93	.336
	Visual memory–del	.0003	1, 258	0.08	.778
	Stroop word	.024	1, 258	6.29	.013
	Stroop color	.023	1, 258	6.14	.014
	Cholesterol	.0008	1, 258	0.20	.654
	Triglycerides	.005	1, 258	1.38	.241

Note: PCBs = log transformed PCB exposure, A = age, G = gender, imm = immediate recall, del = delayed recall.

Testing Multivariate Hypotheses on Sets of Predictors and Other Complex Hypotheses

Occasions often arise in whic]h the question of interest about a multivariate set of observations is whether a set of two or more predictors, adjusted for the remaining predictors, can account for a significant proportion of the joint variance in the dependent variables. Such tests usually follow from a logical analysis of a question in which a set of predictors taken together identifies a higher order construct; for example, income, education, and occupational prestige can be taken as indicators of socioeconomic status (SES), but SES cannot be defined solely with reference to any single predictor. A model formulated to test the construct SES might usefully estimate the joint effect of the set of three predictors, adjusted for any remaining covariates in the model.

Consider an example based on the personality data in which it is of interest to know if the set of positive personality variables (X_2 = Extraversion and X_3 = Conscientiousness) are significantly related to the four job search response variables. The null hypothesis that requires two rows in $\mathbf{L}(q_h = q_f - q_r = 2)$ is

$$H_0:\mathbf{LB} = \begin{bmatrix} 0 & 0 & 1 & 0 \\ 0 & 0 & 0 & 1 \end{bmatrix} \begin{bmatrix} \beta_{01} & \beta_{02} & \beta_{03} & \beta_{04} \\ \beta_{11} & \beta_{12} & \beta_{13} & \beta_{14} \\ \beta_{21} & \beta_{22} & \beta_{23} & \beta_{24} \\ \beta_{31} & \beta_{32} & \beta_{33} & \beta_{34} \end{bmatrix}$$

$$= \begin{bmatrix} \beta_{21} & \beta_{22} & \beta_{23} & \beta_{24} \\ \beta_{31} & \beta_{32} & \beta_{33} & \beta_{34} \end{bmatrix} = \begin{bmatrix} 0 & 0 & 0 & 0 \\ 0 & 0 & 0 & 0 \end{bmatrix}.$$

Substituting $\hat{\mathbf{B}}$ for $\mathbf{B}$ in Equation 5.1, the resulting hypothesis SSCP matrix for Extraversion and Conscientiousness adjusted for Neuroticism is

$$\mathbf{Q}_{H_{(E,C\,|\,N)}} = \begin{bmatrix} 134.123 & 11.089 & 13.020 & -6.355 \\ 11.089 & 358.655 & 28.079 & 18.859 \\ 13.020 & 28.079 & 3.302 & .846 \\ -6.355 & 18.859 & .846 & 1.351 \end{bmatrix}.$$

Using $\mathbf{Q}_E$ from Table 4.1, the four multivariate test statistics, R_m^2s, and F-test approximations are collected in Table 5.5. With $s = \text{minimum}[p, q_h]$ = 2, the F-tests for Pillai's, Hotelling's, and Roy's criteria are approximate; the F test for Wilks' criterion is exact for $s = 2$. The last line of the table is the SAS PROC GLM exact test for Roy's criterion. Note the discrepancy between the F and p values for all four test statistics as they do not provide the same information when $s \neq 1$; it matters little in this example because of the strength of the result. Olson (1974, 1976) gives advice on the relative power differentials among the four test statistics.

The univariate follow-up tests of Table 5.6 suggest that the multivariate relationship between this pair of positive predictors is dependent on all four

Table 5.5 Multivariate Test Statistics on the Set of Variables (E, C | N)

Hypothesis	*Test*	*Statistic*	R^2_m	ν_h, ν_e	*F-Approx.*	*p*
E, C | N	Pillai's V	.425	.213	8, 186	6.28	.0000003
	Wilks' Λ	.619	.213	8, 184	6.22	.0000004
	Hotelling's T	.543	.213	8, 182	6.17	.0000005
	Roy's θ	.238	.238	4, 93	7.25	.0000397
	Roy's θ (exact)	.238	.238	4, 93	—	.0002000

Note: *F*-test for Λ is exact. Exact test for θ is the MSTAT=EXACT option of SAS PROC REG. N = Neuroticism, E = Extraversion, C = Conscientiousness.

criteria, but these follow-up tests take no account of the relationships between response variables and may therefore be overestimates.

Reference to the individual partialled results of Tables 5.1 and 5.2 reveals that the results of this test on the Extraversion + Conscientiousness set subsumes most of the results observed for the individual tests. In this example, there would be no compelling theoretical reason to report the test of this set of predictors in lieu of the individual predictors. In studies in which the higher order construct is of primary theoretical interest (e.g., the SES example), then tests of sets may be the test of choice.

Testing Other Complex Multivariate Hypotheses

One advantage of the general linear hypothesis test is the ease with which more complex hypotheses can be formulated. These complex hypotheses will make the most sense when based on strong a priori theoretical explanations of the anticipated relationship. As an example, consider the PCB data in which age and exposure to PCBs are two competing explanations for the observed psychological and physiological dysfunction. In previous analyses, each was found to be a significant predictor of the response variables but that does not address the question of superiority of one predictor variable over the other. Tests of equality of regression coefficients are best performed on the standard scores of the variables such that the difference between predictors will not be influenced by underlying differences in scale but are tests of differences in correlations.[10] A test of the equality of the age and PCB exposure factors can be specified in a null hypothesis formed by the contrast matrix $\mathbf{L}$ ($q_h = 1$) and the (3 × 6) matrix of parameters $\mathbf{B}^*$ of the standard score full model,

$$
\begin{aligned}
H_0{:}\mathbf{LB}^* &= \begin{bmatrix} 1 & 0 & -1 \end{bmatrix} \begin{bmatrix} \beta^*_{11} & \beta^*_{12} & \beta^*_{13} & \beta^*_{14} & \beta^*_{15} & \beta^*_{16} \\ \beta^*_{21} & \beta^*_{22} & \beta^*_{23} & \beta^*_{24} & \beta^*_{25} & \beta^*_{26} \\ \beta^*_{31} & \beta^*_{32} & \beta^*_{33} & \beta^*_{34} & \beta^*_{35} & \beta^*_{36} \end{bmatrix} \\
&= \begin{bmatrix} \beta^*_{11}-\beta^*_{31} & \beta^*_{12}-\beta^*_{32} & \beta^*_{13}-\beta^*_{33} & \beta^*_{14}-\beta^*_{34} & \beta^*_{15}-\beta^*_{35} & \beta^*_{16}-\beta^*_{36} \end{bmatrix} \\
&= \begin{bmatrix} 0 & 0 & 0 & 0 & 0 & 0 \end{bmatrix},
\end{aligned}
$$

which is a multivariate hypothesis concerning six sets of differences between standardized regression coefficients. Substituting $\hat{\mathbf{B}}^* = \mathbf{R}^{-1}_{XX}\mathbf{R}_{XY}$ (Equation 3.11) for $\mathbf{B}^*$ of the hypothesis and evaluating Equation 5.13 for the model based on standard scores leads to the desired hypothesis SSCP matrix,

$$
\mathbf{Q}^*_H = \left(\mathbf{L}\hat{\mathbf{B}}^*\right)'\left(\mathbf{L}\mathbf{R}^{-1}_{XX}\mathbf{L}'\right)^{-1}(\mathbf{L}\hat{\mathbf{B}}^*), \qquad [5.13]
$$

---

[10]The test of the differences between two row vectors of standardized coefficients, say $\boldsymbol{\beta}^*_{Y\cdot X_1}$ versus $\boldsymbol{\beta}^*_{Y\cdot X_2}$, is a test of the differences between multivariate semipartial correlations and is conceptually quite different from a test of differences in raw regression coefficients. The scales of age and log-transformed PCBs are so different as to render any comparison of unstandardized coefficients difficult to interpret. [See footnote 18, Chapter 1]

Table 5.6 Univariate Follow-Up Tests for (E, C | N)

Hypothesis	*Response Variable*	R^2	ν_h, ν_e	F	p
E, C | N	Background Prep.	.099	2, 95	4.26	.017
	Social Prep.	.013	2, 95	8.30	<.001
	Follow-up Interview	.104	2, 95	9.51	<.001
	Offers	.046	2, 95	6.34	.003

Note: N = Neuroticism; E = Extraversion; C = Conscientiousness.

Table 5.7 Multivariate Test Statistics on the Difference Model (Age - PCBs | Gender) for the PCB Data

Hypothesis	*Test*	*Statistic*	R^2_m	ν_h, ν_e	*F-Approx.*	*p*
A – PCB | G	Pillai's *V*	.016	.016	6, 253	.70	.650
	Wilks' Λ	.984	.016	6, 253	.70	.650
	Hotelling's T	.017	.016	6, 253	.70	.650
	Roy's θ	.017	.016	6, 253	.70	.650

Note: All *F*-tests are exact and equivalent for s = 1. PCBs = log transformed PCB exposure, A = age, G = gender.

we find

$$\mathbf{Q}^*_{H_{(age-PCBs \mid gender)}} = \begin{bmatrix} .0003 & .0001 & .0013 & .0017 & .0005 & .0010 \\ .0001 & .0000 & .0004 & .0006 & .0002 & .0004 \\ .0013 & .0004 & .0058 & .0079 & .0023 & .0047 \\ .0017 & .0006 & .0079 & .0109 & .0032 & .0065 \\ .0005 & .0002 & .0023 & .0032 & .0009 & .0019 \\ .0010 & .0004 & .0047 & .0065 & .0019 & .0038 \end{bmatrix}.$$

The error SSCP for standard scores is found by Equation 3.9,

$$\mathbf{Q}^*_{E_{(age-PCBs \mid gender)}} = \begin{bmatrix} .8332 & .6178 & .1288 & .0764 & .0461 & .0789 \\ .6178 & .8387 & .1265 & .0898 & .0161 & .0548 \\ .1288 & .1265 & .9363 & .6856 & -.0308 & .0386 \\ .0764 & .0898 & .6586 & .9103 & -.0491 & .0206 \\ .0461 & .0161 & -.0308 & -.0491 & .8419 & .4089 \\ .0789 & .0548 & .0386 & .0206 & .4089 & .8418 \end{bmatrix}.$$

Using $\mathbf{Q}^*_H$ and $\mathbf{Q}^*_E$, the multivariate test statistics can be evaluated by the definitions of Table 4.5 and are summarized in Table 5.7.

There is no evidence in these data that PCBs make a more substantial contribution than age in predicting the joint variance accounted for in the six psychological and physiological response variables. While both individual effects uniquely account for significant proportions of the joint variation in the six response variables (Tables 5.3 and 5.4), one is not significantly superior to the other. Further examples of more complex equality tests of this type are discussed by Rencher (1998, pp. 272, 297) in the multivariate case and Rindskopf (1984) for the univariate linear model.

The Assumptions That Apply to All Multivariate Linear Model Analyses

In Chapter 3, the estimated parameters $\hat{\mathbf{B}}$ of the linear model were seen to be optimal if the assumptions of linearity, constant variance, and independence could be met. In addition to these assumptions, a fourth assumption about the multivariate normal distribution of the errors, $\boldsymbol{\epsilon}_i \sim N_p(\mathbf{0}, \boldsymbol{\Sigma})$, or equivalently $\mathbf{y}_i \sim N_p(\mathbf{XB}, \boldsymbol{\Sigma})$, was required to justify the validity of the test statistics and their approximate F tests. We do not pursue issues of diagnosing the adequacy of the models fitted here with respect to these assumptions but such methods are described in Rencher (1998, Chap. 7) and Stevens (2007, Chap. 5).

In Chapter 6, we expand the scope of multivariate linear model analysis to cases where the design matrix $\mathbf{X}$ is coded in such a way as to represent group membership—the classical definition of the multivariate analysis of variance.

CHAPTER 6. CODING THE DESIGN MATRIX AND MULTIVARIATE ANALYSIS OF VARIANCE

The univariate analysis of variance (ANOVA) is a special case of the general linear model and can be analyzed by regression analysis (Cohen et al., 2003) if the categorical predictor variables denoting group membership are properly coded. Multivariate analysis of variance (MANOVA) is also a special case of the multivariate multiple regression model (MMR); both MANOVA and MMR are themselves special cases of the multivariate general linear model (Tatsuoka, 1988, Chap. 9). The linear model $\mathbf{Y}=\mathbf{XB}+\mathbf{E}$ follows the same rules in both MANOVA and MMR. The major characteristic that distinguishes these two forms of the multivariate linear model is the structure and type of variables contained in the design matrix $\mathbf{X}$. The MMR model is usually characterized by predictor variables that are continuously distributed (quantitative); these predictor *variables* can be represented by a single *vector* of scores in the design matrix. On the other hand, variables that represent group membership, consisting of two or more nominal groups, are design features that characterize the ANOVA. In ANOVA designs, we distinguish between the *variable* or factor (say, group) and the *vector* (or vectors) that are required to represent the variable in the design matrix. For the typical ANOVA factor, there is often no underlying metric on which the groups can be ordered—the allocation of group membership is nominal and arbitrary and some sort of coding scheme is required to distinguish between groups and to represent the information contained in group differences. Analyzing multiple dependent variables in the context of categorical predictor variables defines the MANOVA. Once a coding scheme is adopted for a categorical ANOVA factor, the solution to the multivariate linear model problem for MANOVA follows the same pattern as presented in the previous chapters. To make matters more concrete, we will first introduce the mechanics of coding categorical variables in a two-group design and discuss the differences in the meaning of parameter estimates that accompany different coding schemes, and we will then extend the rules for coding ANOVA factors to designs with more than two groups. We will illustrate the solution to two MANOVA example problems—a four-group, one-way classification MANOVA and a 3 × 2 factorial MANOVA, each with three dependent variables.

The Variable Versus Vector Distinction

When constructing the design matrix for ANOVA-type problems that we illustrate here, it is imperative to make a distinction between the labels "variable" and "vector." If the design matrix **X** contains only continuously distributed, quantitative measures, as has been the case in the previous chapters, the variable and the single vector it takes to represent that variable in **X** are one and the same. However, when the *variable* is a categorical, qualitative designation of more than two groups, say Treatments A, B, C, and Control, then the variable of group membership of the individual cases cannot be captured in a single vector of scores. If we let a single vector $\mathbf{x} = 1, 2, 3$, or 4 to identify group membership, for example, the numbers in the vector are arbitrarily assigned and fitting the model $\mathbf{Y} = \mathbf{XB} + \mathbf{E}$ to a vector $\mathbf{X}_{(n\times 1)}$ of arbitrary group identification numbers has no meaning. Systems of dummy variable coding have been devised to avoid this problem. As a general rule, it will take $G - 1$ *vectors* to code the information in a *variable* that is defined by G groups. If G is the number of levels (groups) in the categorical ANOVA factor, then G vectors plus an intercept in **X** defines an *overparameterized* model that is intuitively sensible but that has no unique solution to the parameter estimates. The many coding schemes that are available are designed to *reparameterize* the model such that a unique solution to the parameter estimates can be obtained, and the analyst must choose one such scheme. The differences in the meaning of the resulting parameter estimates derive from the choice of coding scheme. We begin by introducing dummy coding for the overparameterized model and for two different reparameterized models for categorical variables in two- and three-group ANOVA designs. Following this pedagogical introduction, we illustrate the linear model analysis with a four-group, single-factor MANOVA with $p = 3$ response variables.

Representing a Categorical Variable by Coded Vectors

There are several methods for coding design matrices in ANOVA problems, including overparameterized less-than-full-rank coding, reference cell dummy coding, cell mean coding, effects coding, polynomial coding, and even nonsense coding. Each of these methods has certain unique characteristics, and they are given extensive coverage in Cohen et al. (2003, Chap. 8) and Muller and Fetterman (2002, Chap. 12). For our purposes, three of these coding schemes will suffice: the overparameterized model, the reference cell model, and the cell means model.

The Overparameterized, Less-Than-Full-Rank, Coding Scheme

Assume that we have a two-group ANOVA design with $G = 2$ levels of the factor represented by a treatment and a control group.[1] A model with $p = 2$ response variables and n observations composed of n_1 cases in the first group and n_2 cases in the second group can be represented by a coding scheme for **X** that includes a unit vector, $X_0 \equiv 1$, and two vectors (X_1, X_2) of 1s and 0s to identify group membership. The multivariate linear regression model for this ANOVA design would be defined as

$$\mathbf{Y}_{(n\times p)} = \mathbf{X}_{(n\times q+1)}\mathbf{B}_{(q+1\times p)} + \mathbf{E}_{(n\times p)}. \qquad [6.1]$$

Setting $n_1 = n_2 = 2$, an example setup of the model with a unit vector and two vectors of 1-0 dummy variables would be

$$\begin{bmatrix} Y_{11}^{(1)} & Y_{12}^{(1)} \\ Y_{21}^{(1)} & Y_{22}^{(1)} \\ Y_{31}^{(2)} & Y_{32}^{(2)} \\ Y_{41}^{(2)} & Y_{42}^{(2)} \end{bmatrix} = \begin{bmatrix} 1 & 1 & 0 \\ 1 & 1 & 0 \\ 1 & 0 & 1 \\ 1 & 0 & 1 \end{bmatrix} \begin{bmatrix} \beta_{01} & \beta_{02} \\ \beta_{11} & \beta_{12} \\ \beta_{21} & \beta_{22} \end{bmatrix} + \begin{bmatrix} \varepsilon_{11} & \varepsilon_{12} \\ \varepsilon_{21} & \varepsilon_{22} \\ \varepsilon_{31} & \varepsilon_{32} \\ \varepsilon_{41} & \varepsilon_{42} \end{bmatrix},$$

where the superscript, $g = 1, 2, \ldots, G$, in the notation $Y_{ik}^{(g)}$ denotes group membership and the subscripts define the subject and response variable. Focus attention on the design matrix **X** with elements X_{ij} denoting the ith case on the jth predictor variable, $i = 1$ to n, $j = 1$ to q. For the $n_1 + n_2 = 4$ cases, each row of **X** identifies a case's location in the group structure denoted by the superscript of rows of **Y**. Recall from Chapter 3 (Equation 3.4) that the estimate $\hat{\mathbf{B}}$ of $\mathbf{B}$, given by

$$\hat{\mathbf{B}} = (\mathbf{X}'\mathbf{X})^{-1}\mathbf{X}'\mathbf{Y}, \qquad [6.2]$$

[1]The two groups need not be from a designed experiment but can be any two qualitatively distinct groups. If the groups are formed by true experiment, then the allowable inferences from the analysis would be different than if the groups are observational in nature, but the procedures for analyzing both experimental and observational designs would be the same.

requires the inverse of $\mathbf{X'X}$ as a first step in the solution. For these prototype data, $\mathbf{X'X}$ is

$$\mathbf{X'X} = \begin{bmatrix} 1 & 1 & 1 & 1 \\ 1 & 1 & 0 & 0 \\ 0 & 0 & 1 & 1 \end{bmatrix} \begin{bmatrix} 1 & 1 & 0 \\ 1 & 1 & 0 \\ 1 & 0 & 1 \\ 1 & 0 & 1 \end{bmatrix} = \begin{bmatrix} n & n_1 & n_2 \\ n_1 & n_1 & 0 \\ n_2 & 0 & n_2 \end{bmatrix}.$$

To take the inverse of $\mathbf{X'X}$, it is necessary to evaluate its determinant (see Fox, 2009, Sect. 1.1.4). Observing that $n = n_1 + n_2$, the determinant of this dummy-coded design matrix can be shown to be zero,

$$|\mathbf{X'X}| = \left[\left(nn_1n_2 + n_1 0 n_2 + n_2 n_1 0 \right) - \left(n_1 n_1 n_2 + n00 + n_2 n_1 n_2 \right) \right] = 0.$$

Since the inverse of this matrix requires division by the determinant, then $\mathbf{X'X}$ is singular and $(\mathbf{X'X})^{-1}$ does not exist for these data. Consequently, $\hat{\mathbf{B}}$ cannot be uniquely estimated. A matrix of full rank is one in which the columns are linearly independent of one another, whereas a less-than-full-rank matrix contains columns that are a linear combination of other columns in the matrix. A determinant of a matrix equal to zero is diagnostic of linear dependence in a less-than-full-rank matrix. Note that the first column of $\mathbf{X}$ in the example above is the sum of the second and third columns; the same is true of $\mathbf{X'X}$. Hence, both $\mathbf{X}$ and $\mathbf{X'X}$ are said to be of deficient rank, with determinant of 0 and no inverse. It is easy to see that this *overparameterized* design matrix $\mathbf{X}$ contains too much information; knowing X_0 and X_1 is sufficient to identify all members of both groups in the ANOVA design. This principle applies to all ANOVA factors in which the number of vectors exceeds the number or levels (groups) of the factor. A three-group, single-factor ANOVA design coded with vectors X_0, X_1, X_2, and X_3 would also be of deficient rank and have no unique inverse of $\mathbf{X'X}$.

There are two general solutions to this problem: (1) a generalized inverse (see Muller & Fetterman, 2002, Appendix A.2), which is the method implemented in most commercial software for multivariate analysis, or (2) *reparameterizing* the model in such a way as to remove the linear dependence and obtain a unique solution to $(\mathbf{X'X})^{-1}$. This latter solution, which is intuitively appealing, is achieved by altering the overparameterized coding scheme displayed in $\mathbf{X}$ above. Among the several available schemes for reparameterized models, we restrict our attention to the reference cell model and the cell means model. They are both full-rank methods, produce interpretable parameter estimates, and are easily compatible with the general linear tests associated with the contrasts $\mathbf{L}\hat{\mathbf{B}}=0$.[2]

The Reference Cell Dummy Coding Scheme

The reference cell reparameterized model retains the use the 1s and 0s of the overparamaterized model but deletes the rightmost column of the design matrix. Deleting this column leaves **X** of full rank and makes it possible to obtain a unique inverse of $\mathbf{X'X}$. The general pattern of reference cell dummy variable coding the 1s and 0s is as follows for any number of groups, where coding stops with $q = G-1$ vectors.

Let $X_1 = 1$, if the participant is from Group 1,
otherwise let $X_1 = 0$.
Let $X_2 = 1$, if the participant is from Group 2,
otherwise let $X_2 = 0$.
Let $X_3 = 1$, if the participant is from Group 3,
otherwise let $X_3 = 0$.

⋮

Let $X_q = 1$, if the participant is from Group $G-1$,
otherwise let $X_q = 0$.

The coding schemes for **X** for two-group, three-group, and four-group single-classification ANOVA designs are shown in Table 6.1, where the number of columns (in addition to X_0) is always equal to $G-1$, or one less than the number of groups in the factor.

The term reference cell coding derives from the fact that group coded as 0 on all vectors other than X_0 serves as the reference group. The practical implications of this coding scheme are seen in the meaning of the parameter estimates of Equation 6.1 that result from reference cell coding. Using the two-group, $n = 4$ example setup with design matrix **X** from Table 6.1 in

[2]The parameter estimates associated with reference cell coding are the estimates printed by GLM procedures in SAS, SPSS, STATA, and other commercial software even though these programs employ the generalized inverse to obtain the solution.

Table 6.1 Reference Cell Coding Scheme for Two-, Three-, and Four-Group One-Way ANOVA Designs

2 Group	*3 Group*	*4 Group*
$\begin{bmatrix} 1 & 1 \\ 1 & 1 \\ 1 & 0 \\ 1 & 0 \end{bmatrix}$	$\begin{bmatrix} 1 & 1 & 0 \\ 1 & 1 & 0 \\ 1 & 0 & 1 \\ 1 & 0 & 1 \\ 1 & 0 & 0 \\ 1 & 0 & 0 \end{bmatrix}$	$\begin{bmatrix} 1 & 1 & 0 & 0 \\ 1 & 1 & 0 & 0 \\ 1 & 0 & 1 & 0 \\ 1 & 0 & 1 & 0 \\ 1 & 0 & 0 & 1 \\ 1 & 0 & 0 & 1 \\ 1 & 0 & 0 & 0 \\ 1 & 0 & 0 & 0 \end{bmatrix}$

which Group 2 is the reference cell, $\mathbf{X'X} = \begin{bmatrix} n & n_1 \\ n_1 & n_1 \end{bmatrix}$. Assuming $p=2$ response variables, the parameter estimates can be shown to be

$$\hat{\mathbf{B}}_{(q+1\times p)} = (\mathbf{X'X})^{-1}\,\mathbf{X'Y} = \begin{bmatrix} \dfrac{1}{n_1 n_2} & -\dfrac{1}{n_1 n_2} \\ -\dfrac{1}{n_1 n_2} & \dfrac{n_1 + n_2}{n_1 n_2} \end{bmatrix} \begin{bmatrix} \Sigma Y_1 & \Sigma Y_2 \\ \Sigma Y_1^{(1)} & \Sigma Y_2^{(1)} \end{bmatrix}$$

$$= \begin{bmatrix} \hat{\beta}_{01} & \hat{\beta}_{02} \\ \hat{\beta}_{11} & \hat{\beta}_{12} \end{bmatrix} = \begin{bmatrix} \overline{Y}_1^{(2)} & \overline{Y}_2^{(2)} \\ \overline{Y}_1^{(1)} - \overline{Y}_1^{(2)} & \overline{Y}_2^{(1)} - \overline{Y}_2^{(2)} \end{bmatrix}.$$

For this $p = 2$ example, the estimated parameters have a known regular pattern; the intercepts $\hat{\beta}_{01}$ and $\hat{\beta}_{02}$ are always the means of Y_1 and Y_2 for the group coded as the reference cell (Group 2 in this example). The estimates $\hat{\beta}_{11}$ and $\hat{\beta}_{12}$ are the differences between the means of the group coded 1 on the vector X_1 and the means of the reference cell group. ANOVA designs with three, four, or more groups have similar structures for the parameter estimates; $\hat{\beta}_{01}, \hat{\beta}_{02}, \cdots, \hat{\beta}_{0p}$ will contain the means of the reference cell, while the remaining estimates will define a contrast between the means of the group coded 1 on vector X_j and the mean of the reference cell. In multivariate designs, these mean differences will be duplicated for all p dependent variables. Applying the results of Chapters 3 to 5 to this matrix of parameter estimates leads to a series of tests on the individual parameter estimates, the multivariate whole model estimates, and other possible contrasts on the $\hat{\beta}_{jk}$ values. These tests are illustrated with a four-group, single-classification MANOVA on the stature data that was introduced in Chapter 2, Tables 2.3 and 2.4.

Cell Mean Coding Scheme

The traditional view of ANOVA models is one that focuses on testing of mean differences. From the regression/linear model perspective, the cell mean coding scheme for ANOVA designs involves a simple modification to the overparameterized model in which the parameter estimates $\hat{\mathbf{B}}$ are the cell means of the ANOVA design. Hence the procedures developed for analysis of multivariate linear regression models apply directly to formulating hypotheses about mean differences when cell mean coding is used. To illustrate, assume that we have an overparameterized three-group, single-classification model with $n_1 = n_2 = n_3 = 2$ and $n = 6$ observations as shown in Table 6.2. Cell mean coding is accomplished by deleting the unit vector X_0 from the overparameterized model.

An advantage of this full-rank reparameterized model is that both

$$\mathbf{X'X} = \begin{bmatrix} n_1 & 0 & 0 \\ 0 & n_2 & 0 \\ 0 & 0 & n_3 \end{bmatrix}$$

Table 6.2 Cell Mean Coding for a Three-Group, One-Way ANOVA Design

3-Group Overparameterized		*3-Group Cell Mean Coded*
$\begin{bmatrix} 1 & 1 & 0 & 0 \\ 1 & 1 & 0 & 0 \\ 1 & 0 & 1 & 0 \\ 1 & 0 & 1 & 0 \\ 1 & 0 & 0 & 1 \\ 1 & 0 & 0 & 1 \end{bmatrix}$	$\rightarrow$	$\begin{bmatrix} 1 & 0 & 0 \\ 1 & 0 & 0 \\ 0 & 1 & 0 \\ 0 & 1 & 0 \\ 0 & 0 & 1 \\ 0 & 0 & 1 \end{bmatrix}$

and

$$(\mathbf{X'X})^{-1} = \begin{bmatrix} \frac{1}{n_1} & 0 & 0 \\ 0 & \frac{1}{n_2} & 0 \\ 0 & 0 & \frac{1}{n_3} \end{bmatrix}$$

are relatively simple.[3] The solution to a three-group linear model with $p = 2$ response variables and a cell mean coding scheme with no intercept would give the parameter estimates as

$$\hat{\mathbf{B}}_{(q \times p)} = (\mathbf{X'X})^{-1}\mathbf{X'Y} = \begin{bmatrix} \frac{1}{n_1} & 0 & 0 \\ 0 & \frac{1}{n_2} & 0 \\ 0 & 0 & \frac{1}{n_3} \end{bmatrix} \begin{bmatrix} \Sigma Y_1^{(1)} & \Sigma Y_2^{(1)} \\ \Sigma Y_1^{(2)} & \Sigma Y_2^{(2)} \\ \Sigma Y_1^{(3)} & \Sigma Y_2^{(3)} \end{bmatrix} = \begin{bmatrix} \hat{\beta}_{11} & \hat{\beta}_{12} \\ \hat{\beta}_{21} & \hat{\beta}_{22} \\ \hat{\beta}_{31} & \hat{\beta}_{32} \end{bmatrix}$$

$$= \begin{bmatrix} \overline{Y}_1^{(1)} & \overline{Y}_2^{(1)} \\ \overline{Y}_1^{(2)} & \overline{Y}_2^{(2)} \\ \overline{Y}_1^{(3)} & \overline{Y}_2^{(3)} \end{bmatrix}.$$

The q parameter estimates take on the straightforward pattern of being the p group means for each of the G groups in the ANOVA factor. As described in Chapters 3 to 5, a test of hypotheses formulated by the contrast $\mathbf{LB} = \mathbf{0}$ will allow for a variety of tests that are often performed in the traditional ANOVA context, that is, tests of mean differences. Cell mean coding has the additional advantage of being able to explicitly formulate and test hypotheses in more complex ANOVA designs. We illustrate cell mean coding and its related tests of hypotheses on the ANOVA examples in a later section.

---

[3]The inverse of a diagonal matrix is the reciprocal of each of the elements on the main diagonal.

Testing MANOVA Hypotheses via the General Linear Test

The conceptual focus of ANOVA designs is on mean differences. In the univariate case, the ANOVA null hypothesis can be specified in terms of the equality of the population means of the G groups,

$$H_0{:}\mu^{(1)}=\mu^{(2)}=\cdots=\mu^{(G)}, \qquad [6.3]$$

in which the population means of Equation 6.3 are estimated from the sample group means, $\bar{Y}^{(1)}, \bar{Y}^{(2)}, \ldots \bar{Y}^{(G)}$. This univariate hypothesis on group differences can also be equivalently stated as a set of simultaneous contrasts of differences between means such as

$$H_0{:}\begin{bmatrix} \mu^{(1)}-\mu^{(G)} \\ \mu^{(2)}-\mu^{(G)} \\ \vdots \\ \mu^{(G-1)}-\mu^{(G)} \end{bmatrix}=\begin{bmatrix} 0 \\ 0 \\ \vdots \\ 0 \end{bmatrix} \qquad [6.4]$$

In Equation 6.4, the number of contrasts is equal to $G-1$, which is both the degrees of freedom for the ANOVA factor and the number of columns of the reparamaterized design matrix **X**. The multivariate statement of these ANOVA hypotheses merely extends to the equivalent matrix expressions that can accommodate the group differences on p dependent variables. That is,

$$H_0{:}\ \boldsymbol{\mu}^{(1)}=\boldsymbol{\mu}^{(2)}=\cdots=\boldsymbol{\mu}^{(G)}, \qquad [6.5]$$

where each $\boldsymbol{\mu}^{(g)}$ represents a vector of p means, all of which are estimated by the corresponding vectors of sample means within the groups, $\bar{\mathbf{Y}}^{(1)}, \bar{\mathbf{Y}}^{(2)}, \cdots, \bar{\mathbf{Y}}^{(G)}$. The equivalent multivariate hypothesis stated in terms of simultaneous contrasts between group means across all p response variables takes on a form that is an expanded version of the univariate hypothesis,

$$H_0{:}\begin{bmatrix} \mu_1^{(1)}-\mu_1^{(G)} & \mu_2^{(1)}-\mu_2^{(G)} & \cdots & \mu_p^{(1)}-\mu_p^{(G)} \\ \mu_1^{(2)}-\mu_1^{(G)} & \mu_2^{(2)}-\mu_2^{(G)} & \cdots & \mu_p^{(2)}-\mu_p^{(G)} \\ \vdots & \vdots & \ddots & \vdots \\ \mu_1^{(G-1)}-\mu_1^{(G)} & \mu_2^{(G-1)}-\mu_2^{(G)} & \cdots & \mu_p^{(G-1)}-\mu_p^{(G)} \end{bmatrix} =\begin{bmatrix} 0 & 0 & \cdots & 0 \\ 0 & 0 & \cdots & 0 \\ \vdots & \vdots & \ddots & \vdots \\ 0 & 0 & \cdots & 0 \end{bmatrix}, \qquad [6.6]$$

where the superscript, $g = 1, 2, \cdots, G$, denotes group number and the subscript identifies the dependent variable.

The hypotheses stated on means in Equations 6.3 to 6.6 can be equated to hypothesis formulated on the parameters in **B** depending on the coding scheme chosen for **X**. As seen above, there is a direct connection between the elements of the estimated paramater matrix $\hat{\mathbf{B}}$ and the sample estimates of the population means ($\bar{\mathbf{Y}}_k^{(g)}$) depending on the choice of either reference cell or the cell mean coding schemes. Hence stating hypotheses in terms of the parameters of the linear model, that is, **B**, can be made equivalent to the same hypotheses stated in terms of population group means. The equivalence is obtained by formulating appropriate linear combinations of the parameters defined in the contrast $\mathbf{LB} = \mathbf{0}$. For a prototypical four-group, one-way MANOVA design, the correspondence of Equations 6.5 and 6.6 to the parameters of the reference cell coding scheme and the cell mean coding scheme with an appropriately chosen **L** matrix are shown in Table 6.3.

The null hypothesis of no mean differences based on the reference cell mean coding scheme and the cell mean coding scheme differ only in their respective contrast (**L**) and parameter (**B**) matrices since both differ depending on the coding scheme chosen.

The reference cell coding includes an intercept (X_0) and defines a reference cell–coded zero throughout the remaining predictors in **X**. As described above, the reference cell–coded intercept parameter always defines the means of the reference group ($\boldsymbol{\beta}_0$) and the remaining parameters ($\boldsymbol{\beta}_1, \boldsymbol{\beta}_2, \cdots, \boldsymbol{\beta}_j$) always define differences between the mean of the reference group and the means of the remaining groups. The **LB** contrast in the reference cell example of Table 6.3 ignores the intercept and defines a hypothesis on the simultaneous differences between the remaining groups—the classic MANOVA test of the factor. We illustrate a comparable four-group, one-way MANOVA test of this hypothesis with the stature estimation data below.

The cell mean coding scheme in Table 6.3 is a model with no intercept, and the parameter estimates ($\boldsymbol{\beta}_1, \boldsymbol{\beta}_2, \cdots, \boldsymbol{\beta}_j$) are equal to each of the cell (group) means. To define a test of mean differences implied in Equations 6.5 and 6.6 for a four-group MANOVA with p dependent variables, the contrast $\mathbf{LB} = \mathbf{0}$ takes on a slightly different form but with the same end result. Note from the cell mean–coded setup of Table 6.3 that each row of **L** (with hypothesis $df = q_h$ = number of rows of **L**) defines a specific contrast between the means of two selected groups, and the collection of row contrasts defines the overall MANOVA test of the factor. The end result is the same as that given by the reference cell–coded setup. Thus, at the level of testing the overall hypothesis of Equation 6.5, the resulting $\mathbf{Q}_H$ and $\mathbf{Q}_E$ SSCP matrices, and all four multivariate test statistics based on that partition, will be identical for both coding schemes. In the one-way design, it is largely a matter of individual taste as to which scheme to choose. The cell

Table 6.3 Hypothesis Equivalents of ANOVA, Reference Cell–, and Cell Mean–Coded Models

---

The MANOVA Hypothesis Expressed in Mean Differences

$$H_0: \boldsymbol{\mu}^{(1)} = \boldsymbol{\mu}^{(2)} = \boldsymbol{\mu}^{(3)} = \boldsymbol{\mu}^{(4)}$$

$$H_0: \begin{bmatrix} \mu_1^{(1)}-\mu_1^{(4)} & \mu_2^{(1)}-\mu_2^{(4)} & \cdots & \mu_p^{(1)}-\mu_p^{(4)} \\ \mu_1^{(2)}-\mu_1^{(4)} & \mu_2^{(2)}-\mu_2^{(4)} & \cdots & \mu_p^{(2)}-\mu_p^{(4)} \cdots \\ \mu_1^{(3)}-\mu_1^{(4)} & \mu_2^{(3)}-\mu_2^{(4)} & \cdots & \mu_p^{(3)}-\mu_p^{(4)} \end{bmatrix} = \begin{bmatrix} 0 & 0 & \cdots & 0 \\ 0 & 0 & \cdots & 0 \\ 0 & 0 & \cdots & 0 \end{bmatrix}$$

The Reference Cell–Coded Parameter Matrix and MANOVA Hypothesis Equivalence

$$H_0: \mathbf{LB} = \begin{bmatrix} 0 & 1 & 0 & 0 \\ 0 & 0 & 1 & 0 \\ 0 & 0 & 0 & 1 \end{bmatrix} \begin{bmatrix} \beta_{01} & \beta_{02} & \cdots & \beta_{0p} \\ \beta_{11} & \beta_{12} & \cdots & \beta_{1p} \\ \beta_{21} & \beta_{22} & \cdots & \beta_{2p} \\ \beta_{31} & \beta_{32} & \cdots & \beta_{3p} \end{bmatrix} = \begin{bmatrix} \beta_{11} & \beta_{12} & \cdots & \beta_{1p} \\ \beta_{21} & \beta_{22} & \cdots & \beta_{2p} \\ \beta_{31} & \beta_{32} & \cdots & \beta_{3p} \end{bmatrix}$$

$$= \begin{bmatrix} \mu_1^{(1)}-\mu_1^{(4)} & \mu_2^{(1)}-\mu_2^{(4)} & \cdots & \mu_3^{(1)}-\mu_3^{(4)} \\ \mu_1^{(2)}-\mu_1^{(4)} & \mu_2^{(2)}-\mu_2^{(4)} & \cdots & \mu_3^{(2)}-\mu_3^{(4)} \\ \mu_1^{(3)}-\mu_1^{(4)} & \mu_2^{(3)}-\mu_2^{(4)} & \cdots & \mu_3^{(3)}-\mu_3^{(4)} \end{bmatrix} = \begin{bmatrix} 0 & 0 & \cdots & 0 \\ 0 & 0 & \cdots & 0 \\ 0 & 0 & \cdots & 0 \end{bmatrix}$$

The Cell Mean–Coded Parameter Matrix and MANOVA Hypothesis Equivalence

$$H_0: \mathbf{LB} = \begin{bmatrix} 1 & 0 & 0 & -1 \\ 0 & 1 & 0 & -1 \\ 0 & 0 & 1 & -1 \end{bmatrix} \begin{bmatrix} \beta_{11} & \beta_{12} & \cdots & \beta_{1p} \\ \beta_{21} & \beta_{22} & \cdots & \beta_{2p} \\ \beta_{31} & \beta_{32} & \cdots & \beta_{3p} \\ \beta_{41} & \beta_{42} & \cdots & \beta_{4p} \end{bmatrix}$$

$$= \begin{bmatrix} 1 & 0 & 0 & -1 \\ 0 & 1 & 0 & -1 \\ 0 & 0 & 1 & -1 \end{bmatrix} \begin{bmatrix} \mu_1^{(1)} & \mu_2^{(1)} & \cdots & \mu_p^{(1)} \\ \mu_1^{(2)} & \mu_2^{(2)} & \cdots & \mu_p^{(2)} \\ \mu_1^{(3)} & \mu_2^{(3)} & \cdots & \mu_p^{(3)} \\ \mu_1^{(4)} & \mu_2^{(4)} & \cdots & \mu_p^{(4)} \end{bmatrix}$$

$$= \begin{bmatrix} \mu_1^{(1)}-\mu_1^{(4)} & \mu_2^{(1)}-\mu_2^{(4)} & \cdots & \mu_3^{(1)}-\mu_3^{(4)} \\ \mu_1^{(2)}-\mu_1^{(4)} & \mu_2^{(2)}-\mu_2^{(4)} & \cdots & \mu_3^{(2)}-\mu_3^{(4)} \\ \mu_1^{(3)}-\mu_1^{(4)} & \mu_2^{(3)}-\mu_2^{(4)} & \cdots & \mu_3^{(3)}-\mu_3^{(4)} \end{bmatrix} = \begin{bmatrix} 0 & 0 & \cdots & 0 \\ 0 & 0 & \cdots & 0 \\ 0 & 0 & \cdots & 0 \end{bmatrix}$$

mean–coded scheme has the advantage of providing easily recognizable parameter estimates and extends readily to higher order designs. The reference cell coding scheme is perhaps the most common scheme and is the default format of the output of MANOVA results used by many popular computer programs for multivariate analysis. Tests of more specific hypothesis on selected mean differences can be implemented in either scheme. We illustrate such tests in the examples to follow.

Partitioning the SSCP Matrices and Testing Hypotheses in MANOVA

The hypothesis of Equation 6.5 on a *p*-variable MANOVA factor is the equivalent of the full model test introduced in Chapters 4 and 5. It is a simultaneous multivariate test of the differences between the vector of means of the levels of the MANOVA factor. From Chapter 5 (Equation 5.1), the hypothesis SSCP matrix based on a contrast matrix defined by $\mathbf{L}$ and the estimates $\hat{\mathbf{B}}$ of the parameters is given by $\mathbf{Q}_H = \left(\mathbf{L}\hat{\mathbf{B}}\right)\left(\mathbf{L}\left(\mathbf{X}'\mathbf{X}\right)^{-1}\mathbf{L}'\right)^{-1}(\mathbf{L}\hat{\mathbf{B}})'$, and the error SSCP matrix $\mathbf{Q}_E$ is found in the usual way by $\mathbf{Y}'\mathbf{Y} - \hat{\mathbf{B}}'\mathbf{X}'\mathbf{Y}$. With $\mathbf{Q}_H$ and $\mathbf{Q}_E$, both of order $(p \times p)$, in hand, the four multivariate test statistics, their measures of strength of association, and their *F*-test approximations can be found by the definitions given in Table 4.5 and Equations 5.3 to 5.12. The procedures for obtaining all the above are the same as those developed for the linear models of previous chapters.

One-Way MANOVA of the Stature Estimation Data

The stature estimation data of Auerbach and Ruff (2010) can be used to illustrate the four-group, one-way MANOVA on $p = 3$ dependent variables. The authors clustered the 75 archeological sites in North America into 11 regions based on cultural and natural boundaries. They further clustered these regions into Arctic, Temperate West, Great Plains, and Temperate East groups. The 11 regions and the 4 clusters are displayed in Figures 6.1a and 6.1b.

The means (±95% confidence interval [CI]) of the three dependent variables by group membership in the four geographic clusters of Figure 6.1b are displayed in the box plots of Figure 6.2 and Table 2.3.

A four-group MANOVA is fitted to these data to test the hypothesis that the population group means of the three response variables of mean stature, mean relative lower limb length, and mean crural index[4] are equal. The MANOVA hypothesis is given in Table 6.3.

---

[4]The number of skeletons measured at each site ranged from 1 to 42. The data used here, presented in Auerbach and Ruff (2010), are the means of both male and female skeletons per site. A total of 145 mean values are distributed across the 11 regions and four geographic clusters.

Figure 6.1 (a) North American Regions and Sites of Excavation; (b) Geographic Clusters: Arctic (*right slanted lines*), Temperate West (*dots, left*), Great Plains (*left slanted lines*), and Temperate East (*dots, right*)

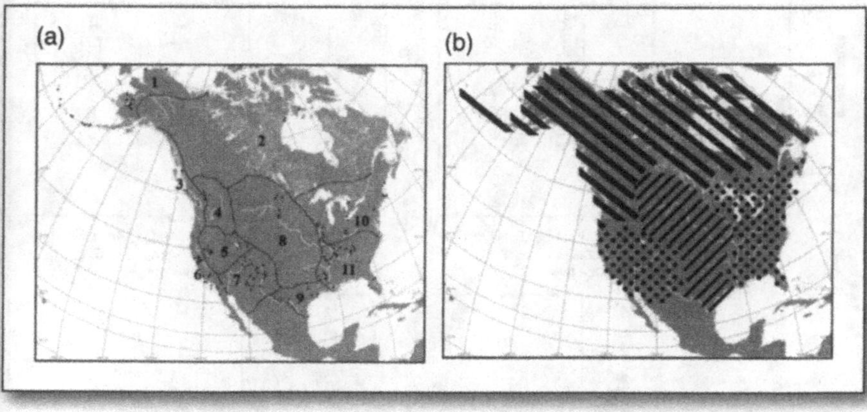

Source: Reproduced from Auerbach and Ruff (2010).

Following the rules for defining the reference cell coding scheme for **X** produces a design matrix with the first column being the unit vector and the remaining $g-1=3$ columns being the group membership–coded dummy variables. The fifth column of the overparameterized design matrix has been deleted to create a full-rank design matrix with a unique inverse of **X′X**. For each group identified in the design matrix below, the row code would be duplicated n_g times. The total sample for this example includes $n = 145$ observations, with the sample sizes for Groups 1 through 4 being 26, 54, 14, and 51, respectively,

$$\mathbf{X}_{(145\times 4)} = \begin{bmatrix} 1 & 1 & 0 & 0 \\ 1 & 0 & 1 & 0 \\ 1 & 0 & 0 & 1 \\ 1 & 0 & 0 & 0 \end{bmatrix} \begin{matrix} \leftarrow \textit{Arctic } (n_1 = 26) \\ \leftarrow \textit{Temperate West } (n_2 = 54) \\ \leftarrow \textit{Great Plains } (n_3 = 14) \\ \leftarrow \textit{Temperate East } (n_4 = 51). \end{matrix}$$

With $p = 3$ response variables, the data matrix **Y** would be of order (145×3). The estimates of the parameters by Equation 6.2 for these data are

$$\hat{\mathbf{B}} = \begin{bmatrix} 161.605 & 49.122 & 84.595 \\ -8.539 & -0.797 & -2.985 \\ -4.406 & -0.478 & 0.285 \\ -0.407 & 0.085 & 1.042 \end{bmatrix}.$$

Figure 6.2 Box Plots of Group Differences on Three Response Variables

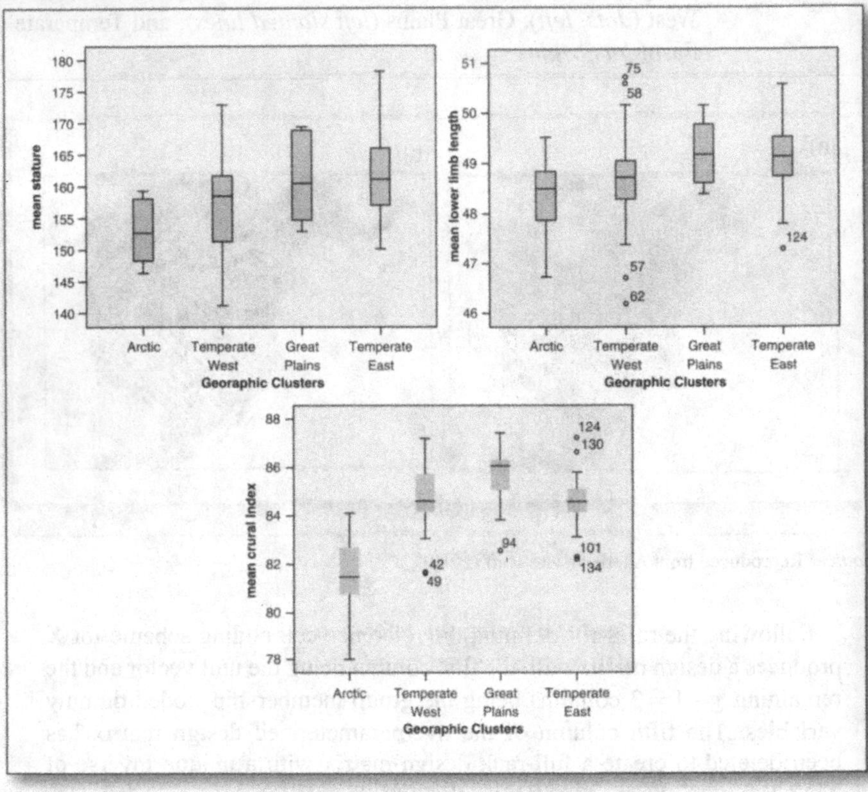

With reference cell coding, the first row of $\widehat{\mathbf{B}}$ contains the sample means of the fourth group (Temperate East) on the response variables of mean stature, lower limb length, and crural index. Rows 2 to 4 of $\widehat{\mathbf{B}}$ contain the successive mean differences of Groups 1, 2, and 3, respectively, versus Group 4 across the three response variables. Interpretation of each of these mean differences (i.e., $\hat{\beta}_{jk}$) might be meaningful if the fourth group has some special status within the MANOVA design. A control group against which three treatments are to be compared would be an example. Otherwise, the choice of reference cell is arbitrary, and the tests on the individual parameters may be uninteresting.

The hypothesis $H_0: \boldsymbol{\mu}^{(1)} = \boldsymbol{\mu}^{(2)} = \boldsymbol{\mu}^{(3)} = \boldsymbol{\mu}^{(4)}$ is evaluated by forming the appropriate contrast matrix written to ignore the intercept and include the remaining parameters in the hypothesis test, that is, $H_0{:}\boldsymbol{\beta}_1 = \boldsymbol{\beta}_2 = \boldsymbol{\beta}_3 = 0$. Thus,

$$\mathbf{L} = \begin{bmatrix} 0 & 1 & 0 & 0 \\ 0 & 0 & 1 & 0 \\ 0 & 0 & 0 & 1 \end{bmatrix}.$$

Estimating $\mathbf{Q}_H$ by Equation 5.1 and the error SSCP matrix $\mathbf{Q}_E$ by Equation 4.3 provides the two matrices necessary to evaluate the whole model and to compute all four multivariate test statistics. These matrices, and their sum, are summarized in Table 6.4.

These three matrices are the constituents necessary to evaluate the four multivariate test statistics, V, Λ, T, and θ as defined in Table 4.5. The evaluation of each test statistic and its measure of strength of association for the four-group MANOVA on the stature data are shown below. The three rows of **L** define $q_h = 3$ and since $p = 3$, the value of $s = \min[p, q_h] = 3$. From Table 4.5, Equations 5.3 to 5.5, and Table 6.4, we find Pillai's Trace V, R_V^2, and the F-test approximation for the hypothesis test on the stature estimation data,

$$V = Tr[(\mathbf{Q}_E + \mathbf{Q}_H)^{-1}\mathbf{Q}_H] = .675,$$

$$R_V^2 = \frac{V}{s} = \frac{.675}{3} = .225,$$

$$F_{V(9,\,423)} = \frac{R_V^2}{1 - R_V^2} \cdot \frac{\nu_e}{\nu_h} = \frac{.225}{1 - .225} \cdot \frac{423}{9} = 13.46, p < .001 \cdot$$

Following the definitions of Table 4.5, Equations 5.6 to 5.12, and the SSCP matrices of Table 6.4, the values of Wilks' Λ, Hotelling's T, and Roy's θ can be found similarly. For Λ we find

$$\Lambda = \frac{|\mathbf{Q}_E|}{|\mathbf{Q}_E + \mathbf{Q}_H|} = \frac{6.87983 * 10^7}{1.74629 * 10^8} = .394,$$

$$R_\Lambda^2 = 1 - \Lambda^{\frac{1}{s}} = 1 - .394^{\frac{1}{3}} = .267,$$

$$F_{\Lambda(9,388.44)} = \frac{R_\Lambda^2}{1 - R_\Lambda^2} \cdot \frac{\nu_e}{\nu_h} = \frac{.267}{1 - .267} \cdot \frac{338.44}{9} = 17.54, p < .001 \cdot$$

Table 6.4 Hypothesis and Error SSCP Matrices for the Multivariate Tests on the MANOVA Factor

$$\mathbf{Q}_H = \begin{bmatrix} 1450.975 & 144.629 & 435.941 \\ 144.629 & 14.787 & 40.864 \\ 435.941 & 40.864 & 235.549 \end{bmatrix}$$

$$\mathbf{Q}_E = \begin{bmatrix} 5852.173 & 325.871 & 196.455 \\ 325.871 & 78.518 & 12.595 \\ 196.455 & 12.595 & 201.367 \end{bmatrix}$$

$$\mathbf{Q}_H + \mathbf{Q}_E = \begin{bmatrix} 7303.149 & 470.500 & 632.396 \\ 470.500 & 93.305 & 53.459 \\ 632.396 & 53.459 & 436.916 \end{bmatrix}$$

And for Hotelling's T we have

$$\mathrm{T} = Tr[\mathbf{Q}_E^{-1}\mathbf{Q}_H] = 1.364,$$

$$R_{\mathrm{T}}^2 = \frac{\mathrm{T}}{\mathrm{T}+s} = \frac{1.364}{1.364+3} = .313,$$

$$F_{\mathrm{T}(9,413)} = \frac{R_{\mathrm{T}}^2}{1-R_{\mathrm{T}}^2} \cdot \frac{\nu_e}{\nu_h} = \frac{.313}{1-.313} \cdot \frac{413}{9} = 20.87, p < .001.$$

As introduced in Chapter 4, Roy's θ is defined as the maximum squared canonical correlation between the **Y** and **X** sets of variables. We discuss the mechanics of obtaining $r_{C_{max}}^2$ in Chapter 7 and for the moment simply report it and its approximate F-test. That is,

$$\theta = r_{C_{max}}^2 = .550,$$

$$F_{\theta(3,141)} = \frac{R_\theta^2}{1-R_\theta^2} \cdot \frac{\nu_e}{\nu_h} = \frac{.550}{1-.550} \cdot \frac{141}{3} = 57.48, p < .001.$$

Based on any of the four multivariate tests, we reject the hypothesis of no difference between the means of four geographic clusters. The proportion of the joint variance in **Y** accounted for by group membership is estimated at anywhere from .23 to .31, depending on which of the first three test statistics one prefers. As shown in previous chapters, the measures of R_m^2 and F take on slightly different numeric values due in part to the inherent differences between arithmetic, geometric, and harmonic means on which these different measures are based. Roy's θ is also notably larger than the remaining three R_m^2 s partly because the joint variance among the variables is not concentrated in a single characteristic root (Olson, 1976). The value of $r_{C_{max}}^2$, therefore, overestimates the joint shared variation between **Y** and group membership. As noted previously, the F-test approximation on Roy's θ is an upper bound and likely to be a liberal overestimate of the p value.[5]

Univariate Follow-Up F Tests on the MANOVA Factor. Univariate follow-up tests were discussed in Chapter 5 (Table 5.2) and have the same informational value in MANOVA designs as they have for multivariate linear models with continuous predictors. Rejection of H_0 on the four-group factor reveals that the vector of means of the three response variables differs across groups. The extent to which each response variable helps discriminate between groups can be evaluated by the univariate R^2s and F-tests based on the diagonal elements of $\mathbf{Q}_H$ and $(\mathbf{Q}_E + \mathbf{Q}_H)$. These univariate quantities are summarized in Table 6.5.

All three of the R^2s are reasonably large by conventional standards (Cohen, 1988) and are statistically significant. All three response variables appear to make a contribution to the group discrimination, keeping in mind that the contribution of each dependent variable is unadjusted for the relationship between the other response variables. The crural index, which is relatively independent of the "size" variables as seen in the correlations of Table 2.4, appears to be the variable that is most influenced by differences between geographic clusters. Since this index is more closely associated with locomotion than with body size, it appears that the multivariate difference, while influenced by all three variables, may be dominated by the crural index. It is known (Auerbach & Ruff, 2010) that the crural index varies systematically with climate and increases as average temperature rises, which suggests that if it is the dominant variable of the three that distinguishes between geographic clusters, there should be a difference between the Arctic region and the samples from warmer climates. Follow-up multivariate hypothesis tests

[5]The exact test of Roy's θ from SAS GLM also provides a value of $p < .001$. The critical value of $\theta^{\alpha}_{(s,m,n)}$ from Harris (2001, table A.5) with $s = 3$, $m = -.5$, and $n = 68.5$ is .098 at $\alpha = .05$, and .126 at $\alpha = .01$.

Table 6.5 Univariate Follow-Up Test Statistics for the MANOVA on Four Geographic Clusters

Hypothesis	*Response Variable*	R^2	df_{h} df_{e}	*F*	*p*
$\mu_1=\mu_2=\mu_3=\mu_4$	Stature	.199	3, 141	11.65	<.001
	Lower limb length	.158	3, 141	8.85	<.001
	Crural index	.539	3, 141	54.98	<.001

of this type, including multivariate pairwise differences, can be tested via the general linear hypothesis we have used throughout.

Multivariate Contrasts on Group Differences. A great many hypotheses about group differences might be formulated following a significant MANOVA, including all $G(G-1)/2 = 6$ pairwise comparisons and many other linear contrasts among the means of the four groups. One advantage of the general linear test approach based on the partition of the SSCP matrices and the general linear test is that among the many possible contrasts, only those that have theoretical substance need to be pursued. The many procedures available for post hoc contrasts in ANOVA designs are complex; extensive discussions can be found in Maxwell and Delaney (2004, Chaps. 4 and 5). In the multivariate case, such contrasts involve both the contrast implemented on the vector of *p* response variables followed by the equivalent univariate contrasts on each of the response variables. Rencher (2002, pp. 182–183) recommends a procedure similar to the Fisher's protected least significant difference procedure for the whole model test and its univariate follow-up *F* tests. If additional contrasts are performed with both multivariate tests and their univariate follow-up tests, some form of error rate protection, such as the Bonferroni correction, should be imposed to avoid unacceptable escalations of Type I error.

Three of the six possible pairwise group contrasts are already available from the reference cell–coded analysis. Examining the multivariate significance of the individual parameter estimates, as defined in the reference cell–coded example of Table 6.3, reveals that a test of the mean differences between Groups 1, 2, or 3 against Group 4 are embodied in the individual parameter estimates of $\hat{\boldsymbol{\beta}}_1$, $\hat{\boldsymbol{\beta}}_2$, and $\hat{\boldsymbol{\beta}}_3$ of the example. To achieve the multivariate test on each of these post hoc contrasts would require three separate $\mathbf{L}$ matrices: $\mathbf{L}_1 = \begin{bmatrix} 0 & 1 & 0 & 0 \end{bmatrix}$, $\mathbf{L}_2 = \begin{bmatrix} 0 & 0 & 1 & 0 \end{bmatrix}$, and $\mathbf{L}_3 = \begin{bmatrix} 0 & 0 & 0 & 1 \end{bmatrix}$. When used separately in conjunction with $\hat{\mathbf{B}}$ and Equation 5.1, three multivariate tests of the mean differences identified by the parameter estimates are performed. The hypothesis SSCP matrices, $\mathbf{Q}_{H_{(1)}}, \mathbf{Q}_{H_{(2)}}$, and $\mathbf{Q}_{H_{(3)}}$ for each of the three multivariate tests are shown in Table 6.6 along with the full model $\mathbf{Q}_E$. The multivariate test statistics are displayed in Table 6.7, and their univariate follow-up F tests are summarized in Table 6.8.

Table 6.6 Hypotheses SSCP Matrices for Contrasts $\mathbf{L}_{(1)} - \mathbf{L}_{(5)}$

$$\mathbf{Q}_{H_{(1)}} = \begin{bmatrix} 1255.569 & 117.245 & 438.882 \\ 117.245 & 10.948 & 40.983 \\ 438.882 & 40.983 & 153.411 \end{bmatrix}$$

$$\mathbf{Q}_{H_{(2)}} = \begin{bmatrix} 509.126 & 55.245 & -32.925 \\ 55.245 & 5.995 & -3.573 \\ -32.925 & -3.573 & 2.129 \end{bmatrix}$$

$$\mathbf{Q}_{H_{(3)}} = \begin{bmatrix} 1.823 & -0.381 & -0.466 \\ -0.381 & 0.080 & 0.975 \\ -4.661 & 0.975 & 11.920 \end{bmatrix}$$

$$\mathbf{Q}_{H_{(4)}} = \begin{bmatrix} 299.774 & 23.158 & 237.156 \\ 23.158 & 1.789 & 18.321 \\ 237.156 & 18.321 & 187.618 \end{bmatrix}$$

$$\mathbf{Q}_{H_{(5)}} = \begin{bmatrix} 949.657 & 91.261 & 469.316 \\ 91.261 & 8.770 & 45.101 \\ 469.316 & 45.101 & 231.933 \end{bmatrix}$$

$$\mathbf{Q}_E = \begin{bmatrix} 5852.173 & 325.871 & 196.455 \\ 325.871 & 78.518 & 12.595 \\ 196.455 & 12.595 & 201.367 \end{bmatrix}$$

Table 6.7 Multivariate Test Statistics on Selected Contrasts of Four Geographic Clusters

Hypothesis	*Test*	*Test Statistic*	R_m^2	ν_h, ν_e	*F-approx.*	*p*
$\boldsymbol{\mu}_1 = \boldsymbol{\mu}_4$	Pillai's *V*	0.469	.469	3, 139	40.94	<.001
	Wilks' Λ	0.531	.469	3, 139	40.94	<.001
	Hotelling's T	0.884	.469	3, 139	40.94	<.001
	Roy's θ	0.469	.469	3, 139	40.94	<.001
$\boldsymbol{\mu}_2 = \boldsymbol{\mu}_4$	Pillai's *V*	0.120	.120	3, 139	6.33	<.001
	Wilks' Λ	0.880	.120	3, 139	6.33	<.001
	Hotelling's T	0.137	.120	3, 139	6.33	<.001
	Roy's θ	0.120	.120	3, 139	6.33	<.001
$\boldsymbol{\mu}_3 = \boldsymbol{\mu}_4$	Pillai's *V*	0.061	.061	3, 139	3.01	.032
	Wilks' Λ	0.939	.061	3, 139	3.01	.032
	Hotelling's T	0.065	.061	3, 139	3.01	.032
	Roy's θ	0.061	.061	3, 139	3.01	.032

Hypothesis	*Test*	*Test Statistic*	R_m^2	ν_h, ν_e	*F-approx.*	*p*
$\boldsymbol{\mu}_1 = \boldsymbol{\mu}_2$	Pillai's *V*	0.483	.483	3, 139	43.35	<.001
	Wilks' Λ	0.517	.483	3, 139	43.35	<.001
	Hotelling's T	0.936	.483	3, 139	43.35	<.001
	Roy's θ	0.483	.483	3, 139	43.35	<.001
$\boldsymbol{\mu}_1 = \frac{\boldsymbol{\mu}_2 + \boldsymbol{\mu}_3 + \boldsymbol{\mu}_4}{3}$	Pillai's *V*	0.549	.549	3,139	56.42	<.001
	Wilks'	0.451	.549	3,139	56.42	<.001
	Hotelling's *T*	1.218	.549	3,139	56.42	<.001
	Roy's θ	0.549	.549	3,139	56.42	<.001

Table 6.8 Univariate Follow-Up Test Statistics on Selected Contrasts of Four Geographic Clusters

Hypothesis	*Response Variable*	R^2	df_h, df_e	F	p
$\boldsymbol{\mu}_1 = \boldsymbol{\mu}_4$	Stature	.177	1, 141	30.25	<.001
	Lower limb length	.122	1, 141	19.66	<.001
	Crural index	.432	1, 141	107.42	<.001
$\boldsymbol{\mu}_2 = \boldsymbol{\mu}_4$	Stature	.080	1, 141	12.27	.001
	Lower limb length	.071	1, 141	10.77	.001
	Crural index	.010	1, 141	1.49	.224
$\boldsymbol{\mu}_3 = \boldsymbol{\mu}_4$	Stature	.000	1, 141	0.04	.834
	Lower limb length	.001	1, 141	0.14	.706
	Crural index	.056	1, 141	8.35	.004
$\boldsymbol{\mu}_1 = \boldsymbol{\mu}_2$	Stature	.049	1, 141	7.22	.008
	Lower limb length	.022	1, 141	3.21	.075
	Crural index	.482	1, 141	131.37	<.001
$\mu_1 = \frac{\mu_2 + \mu_3 + \mu_4}{3}$	Stature	.140	1, 141	22.88	<.001
	Lower limb length	.100	1, 141	15.75	<.001
	Crural index	.535	1, 141	162.40	<.001

The tests on the individual parameter estimates of the reference cell–coded model cover three of the six pairwise comparisons. If the remaining pairwise contrasts are desired, they can be achieved by a contrast based on a hypothesis of differences between parameters. For example, if the difference between the means of Groups 1 and 2 is desired, then the null hypothesis $H_0: \boldsymbol{\beta}_1 - \boldsymbol{\beta}_2 = 0$ implies that the hypothesis $H_0: (\mu_1 - \mu_4) - (\mu_2 - \mu_4) = \mu_1 - \mu_2 = 0$. The **L** matrix required to implement this contrast is $\mathbf{L}_4 = [01-10]$, which along with $\hat{\mathbf{B}}$ and Equation 5.1 leads to $\mathbf{Q}_{H_{(4)}}$ of Table 6.6 and the final tests of Tables 6.7 and 6.8. The remaining pairwise contrasts, if desired, would be computed in like fashion. For the present data, even a Bonferroni split of the critical values (say, $.05/6 = .008$) would not alter the conclusions about statistical significance of the observed differences.

The results summarized in Table 6.7 for the pairwise contrasts that are evaluated in the reference cell–coded model suggest that each of those three contrasts would be declared statistically significant no matter which multivariate test is chosen, although the weakest of the contrasts is between Groups 3 (Great Plains) and 4 (Temperate East). The univariate follow-up tests of this contrast further suggest that Groups 3 and 4 differ mainly on the crural index with little or no contribution from the variables of stature and lower limb length. Conversely, the contrast between Groups 2 (Temperate West) and 4 (Temperate East) suggest that these differences are largely due to the stature and lower limb length (size) with no contribution from the crural index. Although not shown in the table, the multivariate tests on the two remaining pairwise contrasts ($\mu_1 - \mu_3$ and $\mu_2 - \mu_3$) are both statistically significant ($p < .001$) and the univariate follow-up tests show that all three response variables also differ significantly.

A More Complex Contrast. In addition to pairwise comparisons, more complex contrasts can be evaluated by this general linear hypothesis testing strategy. Any estimable hypothesis for which there are unique estimates of the parameters can be tested by these methods.[6] For the stature estimation example data, one estimable hypothesis might focus on the differences between groups that lie on the North-South axis of Figure 6.1, especially with respect to the contribution of the crural index, which is known to vary by climate and by temperature. The hypothesis that the means of the Arctic cluster (Regions 1, 2, 3, and 4 of Figure 6.1a) should

[6]Estimable hypotheses are discussed in detail Green et al. (1999) and Littell, Stroup, and Freund (2002).

differ from the average of the three more temperate geographic clusters can be stated as

$$H_0 : \mu_1 = \frac{\mu_2 + \mu_3 + \mu_4}{3}.$$

This hypothesis can be tested by the contrast vector $\mathbf{L}_{(5)} = [0\ 1 - \frac{1}{3} - \frac{1}{3}]$. When applied to the paramater matrix from the reference cell–coded model, the algebraically equivalent null hypothesis[7] on these mean differences defined by **LB** is given by

$$H_0 : \boldsymbol{\beta}_1 = \frac{1}{3}(\boldsymbol{\beta}_2 + \boldsymbol{\beta}_3).$$

Substituting $\hat{\mathbf{B}}$ and applying Equation 5.1, we find the hypothesis SSCP matrix $\mathbf{Q}_{H_{(5)}}$ of Table 6.6. The multivariate test statistics are shown in Table 6.7, and the univariate follow-up tests are presented in Table 6.8. Any of the multivariate test statistics leads to the rejection of the null hypothesis on this contrast. Clearly, the Arctic region differs significantly from the average of the more temperate climate regions with about 55% of the joint variation of the response variables accounted for by the contrast ($R_V^2 = .549$). Examination of the univariate follow-up F tests leads to the conclusion that this difference is dominated by the crural index ($R^2 = .535$), which is significantly shorter in the Arctic region (81.61) than in the average of the Great Plains and two Temperate clusters (85.03). The stature and lower limb length variables also make statistically significant contributions but with lesser magnitude of effect ($R^2s = .10$ and .14, respectively). The Arctic cluster is of significantly shorter stature and lower limb length (153.07, 48.32) than the average of the remaining clusters (160.00, 48.99). As might be expected from past research, differences in locomotion may be especially sensitive to differences in climate, but differences in size are also detectable in these data.

Higher Order MANOVA Design: A 2 × 3 Factorial MANOVA on the Stature Estimation Data

ANOVA subsumes a large variety of designs beyond the one-way classification, all of which have multivariate realizations (Maxwell & Delaney,

---

[7]Since $\beta_1 = \mu_1 - \mu_4$, $\beta_2 = \mu_2 - \mu_4$, and $\beta_3 = \mu_3 - \mu_4$, $\beta_1 = \frac{1}{3}(\beta_2 + \beta_3)$ is equal to $\mu_1 - \mu_4 = \frac{1}{3}(\mu_2 - \mu_4) + \frac{1}{3}(\mu_3 - \mu_4)$. Adding $\frac{3}{3}\mu_4$ to both sides gives $\mu_1 = \frac{1}{3}\mu_2 - \frac{1}{3}\mu_4 + \frac{1}{3}\mu_3 - \frac{1}{3}\mu_4 + \frac{3}{3}\mu_4$ and $\mu_1 = \frac{1}{3}(\mu_2 + \mu_3 + \mu_4)$.

2004). Among the most widely used is the factorial ANOVA design in which two or more factors can be studied simultaneously. In addition to the main effects of each of the factors, it is also possible to estimate and study the interaction between factors—that is, the extent to which differences in the means of one factor would be conditionally different when classified by the levels of an additional factor.

In the preceding sections, it has been shown that the one-way MANOVA can be solved as a multivariate regression/linear model problem. This was achieved by the device of coding a design matrix **X** to represent the information contained in group membership. It was shown to be important to distinguish between a variable (the ANOVA factor) and the $G - 1$ vectors of **X** required to represent that variable. The regression solution depended on estimating the parameters $\hat{\mathbf{B}}$ of the model, which were shown (Table 6.3) to be either the cell means of the ANOVA groups or some linear combination thereof, depending on the scheme chosen for coding **X**. Finally, tests on hypotheses were accomplished by forming linear contrasts of the parameters, and therefore of the cell means, and evaluating the general linear test by one of the four multivariate test statistics.

This same procedure can be employed to evaluate more complex ANOVA designs that involve more than one factor and may also include the interaction between factors. The whole model test (all main and interaction effects) is seldom of interest in these designs, except for its use in defining the error SSCP matrix; what is of paramount interest are the partialled effects of factors and their interactions.

To make matters more concrete, we focus again on the Auerbach and Ruff (2010) example data presented previously in which the sex of the skeletal cases was noted along with their classification into three clusters based on geographic and cultural similarities. In Table 2.5, the cell means and standard deviations of the 145 observations from 75 excavation sites are organized into a 2×3 factorial MANOVA design. The six cells of the design consist of two levels of sex of skeletons (say, Factor A, $a = 1, \ldots, G_A$) completely crossed with three geographic regions of Arctic, Great Plains, and Temperate groups (say, Factor B, $b = 1, \ldots, G_B$), as illustrated in Figure 6.1b. The cell sample sizes, shown in Table 2.6 along with the response variable means, are not equal, but are roughly proportional.[8] Many

[8]Balanced designs (equal numbers of observations per cell) and unbalanced designs can yield differing solutions to the ANOVA parameter estimates depending on the coding scheme chosen. Imbalance in the cell sample sizes can also introduce confounding among the main effects of the designs. Space limitations preclude a discussion of these issues here, but can be found in Rencher (1998) and Searle (1987). Adopting the cell means coding scheme allows for tests of hypotheses based on the model parameters that are unambiguous. The methods of adjustment for possible confounding of the model effects are discussed in a later section and addressed by the Type I, II, and III methods of partition of the sums of squares in ANOVA and MANOVA designs.

of the design matrix coding schemes mentioned in previous paragraphs lead to parameter estimates in the unbalanced case with contradictory meanings (see Rencher, 1998, Sect. 4.8, for a discussion). In the present illustration, we have adopted the cell means coding scheme introduced in previous sections to implement the design; cell mean coding yields parameter estimates and tests of hypotheses of less ambiguous interpretation and meaning. The means, standard deviations, and cell sample sizes of this 2 (sex) × 3 (geographic cluster) factorial design are presented in Tables 2.5 and 2.6. The multivariate linear model for these $p = 3$ response variables for the 145 observations would be written in the usual linear model form as

$$\mathbf{Y}_{(145\times 3)} = \mathbf{X}_{(145\times 6)}\mathbf{B}_{(6\times 3)} + \mathbf{E}_{(145\times 3)}.$$

The cell means model design matrix would consist of 145 rows by 6 coded vectors; a 2 × 3 factorial yields 6 cell means. Aggregating over the six groups, the design matrix **X** would have a form consistent with the rules defined for cell mean coding—each vector would contain a 1 for any case that is within the $a \times b$ cell and a 0 for any case not a member of that cell of the design. The number of 1s in a cell would be equal to n_{ab}. Thus, the 145 × 6 cell coded design matrix would be constructed as

$$\mathbf{X} = \begin{bmatrix} 1 & 0 & 0 & 0 & 0 & 0 \\ 0 & 1 & 0 & 0 & 0 & 0 \\ 0 & 0 & 1 & 0 & 0 & 0 \\ 0 & 0 & 0 & 1 & 0 & 0 \\ 0 & 0 & 0 & 0 & 1 & 0 \\ 0 & 0 & 0 & 0 & 0 & 1 \end{bmatrix} \begin{matrix} \leftarrow & cell\ a_1b_1, male, Arctic: n_{11} = 13\,rows \\ \leftarrow & cell\ a_1b_2, male, Great\ Plains: n = 55\,rows \\ \leftarrow & cell\ a_1b_3, male, Temperate: n = 7\,rows \\ \leftarrow & cell\ a_2b_1, female, Arctic: n = 13\,rows \\ \leftarrow & cell\ a_2b_2, female, Great\ Plains: n = 50\,rows \\ \leftarrow & cell\ a_2b_3, female, Temperate: n = 7\,rows. \end{matrix}$$

It follows therefore that the product $\mathbf{X}'\mathbf{X}_{(6\times 6)}$ would consist of a diagonal matrix with the value of n_{ab} of Table 2.6 as its elements. The inverse of a diagonal matrix contains the reciprocals of the cell sample sizes on the main diagonal and the necessary $\mathbf{X}'\mathbf{X}^{-1}$ would follow immediately:

$$\mathbf{X}'\mathbf{X} = \begin{bmatrix} n_{11} & 0 & 0 & 0 & 0 & 0 \\ 0 & n_{12} & 0 & 0 & 0 & 0 \\ 0 & 0 & n_{13} & 0 & 0 & 0 \\ 0 & 0 & 0 & n_{21} & 0 & 0 \\ 0 & 0 & 0 & 0 & n_{22} & 0 \\ 0 & 0 & 0 & 0 & 0 & n_{23} \end{bmatrix},$$

$$\mathbf{X'X}^{-1} = \begin{bmatrix} \frac{1}{n_{11}} & 0 & 0 & 0 & 0 & 0 \\ 0 & \frac{1}{n_{12}} & 0 & 0 & 0 & 0 \\ 0 & 0 & \frac{1}{n_{13}} & 0 & 0 & 0 \\ 0 & 0 & 0 & \frac{1}{n_{21}} & 0 & 0 \\ 0 & 0 & 0 & 0 & \frac{1}{n_{22}} & 0 \\ 0 & 0 & 0 & 0 & 0 & \frac{1}{n_{23}} \end{bmatrix}.$$

Substituting $\hat{\mathbf{B}}$ for the parameters of the model equation and using Equation 6.2, the estimated parameters of this 2 × 3 MANOVA model can be found. The product $\mathbf{X'Y}$ of Equation 6.2 are the sums of the Y variables in each of the six cells of the A × B factorial. Hence, the parameter estimates of the fitted linear model of the p response variables are

$$\hat{\mathbf{B}}_{(6\times 6)} = (\mathbf{X'X})^{-1}\mathbf{X'Y} = \begin{bmatrix} \frac{1}{n_{11}} & 0 & 0 & 0 & 0 & 0 \\ 0 & \frac{1}{n_{12}} & 0 & 0 & 0 & 0 \\ 0 & 0 & \frac{1}{n_{13}} & 0 & 0 & 0 \\ 0 & 0 & 0 & \frac{1}{n_{21}} & 0 & 0 \\ 0 & 0 & 0 & 0 & \frac{1}{n_{22}} & 0 \\ 0 & 0 & 0 & 0 & 0 & \frac{1}{n_{23}} \end{bmatrix} \times$$

$$\begin{bmatrix} \Sigma Y_1^{(11)} & \Sigma Y_2^{(11)} & \Sigma Y_3^{(11)} \\ \Sigma Y_1^{(12)} & \Sigma Y_2^{(12)} & \Sigma Y_3^{(12)} \\ \Sigma Y_1^{(13)} & \Sigma Y_2^{(13)} & \Sigma Y_3^{(13)} \\ \Sigma Y_1^{(21)} & \Sigma Y_2^{(21)} & \Sigma Y_2^{(21)} \\ \Sigma Y_1^{(22)} & \Sigma Y_2^{(22)} & \Sigma Y_2^{(22)} \\ \Sigma Y_1^{(23)} & \Sigma Y_2^{(23)} & \Sigma Y_2^{(23)} \end{bmatrix} = \begin{bmatrix} \hat{\beta}_{11} & \hat{\beta}_{12} & \hat{\beta}_{13} \\ \hat{\beta}_{21} & \hat{\beta}_{22} & \hat{\beta}_{23} \\ \hat{\beta}_{31} & \hat{\beta}_{32} & \hat{\beta}_{33} \\ \hat{\beta}_{41} & \hat{\beta}_{42} & \hat{\beta}_{43} \\ \hat{\beta}_{51} & \hat{\beta}_{52} & \hat{\beta}_{53} \\ \hat{\beta}_{61} & \hat{\beta}_{62} & \hat{\beta}_{63} \end{bmatrix}$$

$$= \begin{bmatrix} \bar{Y}_1^{(11)} & \bar{Y}_2^{(11)} & \bar{Y}_3^{(11)} \\ \bar{Y}_1^{(12)} & \bar{Y}_2^{(12)} & \bar{Y}_3^{(11)} \\ \bar{Y}_1^{(13)} & \bar{Y}_2^{(13)} & \bar{Y}_3^{(11)} \\ \bar{Y}_1^{(21)} & \bar{Y}_2^{(21)} & \bar{Y}_3^{(11)} \\ \bar{Y}_1^{(22)} & \bar{Y}_2^{(22)} & \bar{Y}_3^{(11)} \\ \bar{Y}_1^{(23)} & \bar{Y}_2^{(23)} & \bar{Y}_3^{(11)} \end{bmatrix} = \begin{bmatrix} \hat{\mu}_{11} & \hat{\mu}_{12} & \hat{\mu}_{13} \\ \hat{\mu}_{21} & \hat{\mu}_{22} & \hat{\mu}_{23} \\ \hat{\mu}_{31} & \hat{\mu}_{32} & \hat{\mu}_{33} \\ \hat{\mu}_{41} & \hat{\mu}_{42} & \hat{\mu}_{43} \\ \hat{\mu}_{51} & \hat{\mu}_{52} & \hat{\mu}_{53} \\ \hat{\mu}_{61} & \hat{\mu}_{62} & \hat{\mu}_{63} \end{bmatrix}.$$

Consistent with cell mean coding schemes, the parameter estimates are the sample means of the cells of the A × B factorial design, which are also estimates of the population cell means ($\hat{\mu}_{ab}$). The 2 × 3 example contains 6 cells and 6 parameter estimates in the rows of $\hat{\mathbf{B}}$. A 3 × 3 design would contain 9 rows, a 3 × 4 would contain 12 rows, and so on—the product of the factors determines the number of parameter estimates of the fitted linear model. In the matrix $\hat{\mathbf{B}}$ above, the first subscript of the parameter estimates denote the predictor variable (the rows) and the second subscript denotes the dependent variable (the columns). These estimates are matched with the cell means where the parenthesized superscripts denote the row-by-column location within the MANOVA design. The parameter estimates of the cell mean–coded regression model are the sample cell means. The population regression model for this categorical pair of predictor variables (Factors A and B) could also be written as a means model,

$$\mathbf{Y} = \mathbf{X}\boldsymbol{\mu} + \mathbf{E}, \qquad [6.7]$$

which makes explicit the connection between regression parameters ($\hat{\beta}_{ab}$) and the sample means ($\bar{Y}^{(ab)}$)—they are identical estimates of μ_{ab}. Hypotheses in the ANOVA are typically framed in terms of differences in population means; the equivalence of Equations 6.1 and 6.7 also makes explicit the connection between testing mean differences by a contrast **LB** and the partition of $\mathbf{Q}_H$ by way of $\mathbf{L\hat{B}}$. From the definition of the cell mean coding scheme, it is easy to see that evaluating Equation 6.2 on these data yields parameter estimates that are the cell means of Table 2.5. For the Auerbach and Ruff data, the estimates are

$$\hat{\mathbf{B}} = \begin{bmatrix} \hat{\beta}_{11} & \hat{\beta}_{12} & \hat{\beta}_{13} \\ \hat{\beta}_{21} & \hat{\beta}_{22} & \hat{\beta}_{23} \\ \hat{\beta}_{31} & \hat{\beta}_{32} & \hat{\beta}_{33} \\ \hat{\beta}_{41} & \hat{\beta}_{42} & \hat{\beta}_{43} \\ \hat{\beta}_{51} & \hat{\beta}_{52} & \hat{\beta}_{53} \\ \hat{\beta}_{61} & \hat{\beta}_{62} & \hat{\beta}_{63} \end{bmatrix} = \begin{bmatrix} 156.74 & 48.54 & 81.71 \\ 164.04 & 49.11 & 85.02 \\ 167.65 & 49.62 & 85.83 \\ 149.39 & 48.11 & 81.51 \\ 154.17 & 48.62 & 84.44 \\ 154.74 & 48.79 & 85.45 \end{bmatrix},$$

where the columns contain the cell means for stature, lower limb length, and crural index.

The A Main Effect. In this 2 × 3 factorial design, three hypotheses are of interest—the main effect of Factor A, the main effect of Factor B, and the A-by-B interaction. The hypothesis on the A main effect conjectures that the marginal means of males and females are equal in the population. In terms of the cell means, the hypothesis is

$$H_{0(\mathrm{A})}\colon \left(\boldsymbol{\mu}_{a_1b_1} + \boldsymbol{\mu}_{a_1b_2} + \boldsymbol{\mu}_{a_1b_3}\right) = \left(\boldsymbol{\mu}_{a_2b_1} + \boldsymbol{\mu}_{a_2b_2} + \boldsymbol{\mu}_{a_2b_3}\right).$$

A contrast matrix $\mathbf{L}_\mathrm{A}$, with a single row vector such that $q_{h_\mathrm{A}} = 1$, defines this difference as a function of the unweighted averages of the cell means,[9]

$$\mathbf{L}_\mathrm{A} = \frac{1}{3}[1 \quad 1 \quad 1 \quad -1 \quad -1 \quad -1].$$

---

[9]The elements of the contrast vector $\mathbf{L}_\mathrm{A}$ are divided by 3 since the cell means are averaged across three levels of B. The disparities in cell sample sizes are ignored by this method.

The comparable null hypothesis stated as the contrast **LB** would therefore be

$$H_{0(\mathrm{A})}: \frac{1}{3}\mu_{a_1b_1} + \frac{1}{3}\mu_{a_1b_2} + \frac{1}{3}\mu_{a_1b_3} - \frac{1}{3}\mu_{a_2b_1} - \frac{1}{3}\mu_{a_2b_2} - \frac{1}{3}\mu_{a_2b_3} = 0 \cdot$$

Substituting the estimates $\hat{\mathbf{B}}$ into the above and performing the arithmetic yields the 1×3 vector of sample mean differences for males versus females, averaged across the three geographic clusters on the three dependent variables,

$$\mathbf{L}_{\mathrm{A}}\hat{\mathbf{B}} = [10.05 \quad .58 \quad .38] \cdot$$

Using $\hat{\mathbf{B}}$ and the same computational methods used in the previous problems, we estimate $\mathbf{Q}_E = \mathbf{Y}'\mathbf{Y} - \mathbf{B}'\mathbf{X}'\mathbf{Y}$ for this factorial MANOVA and apply Equation 5.1 to obtain the hypothesis SSCP for the A main effect, $\mathbf{Q}_{H(\mathrm{A})}$. Both these SSCP matrices are shown in Table 6.9. Using the definitions of Table 4.5, the multivariate test statistics, R_m^2s, approximate F-tests, and degrees of freedom are computed and are shown in Table 6.10. Degrees of freedom for the multivariate tests are as defined in Equations 5.3 to 5.12—for the single vector in the contrast vector $\mathbf{L}_{\mathrm{A}}$ we have $q_{h\mathrm{A}} = 1$ corresponding to the $G_{\mathrm{A}} - 1 = 2 - 1 = 1$ degree of freedom available for this two-level factor. The multivariate hypothesis degrees of freedom are $\nu_h = p \times q_{h\mathrm{A}} = 3 \times 1 = 3$; the error degrees of freedom are defined separately for each of the four test statistics as defined in Equations 5.3 to 5.12. The univariate follow-up tests have also been presented in Table 6.11.

The results of the analysis of the A main effect clearly indicate that the vector of means of the response variables differ significantly between males and females. The multivariate difference is supported by any of the four multivariate test statistics, revealing that about 41% of the joint variance of the response variables is accounted for by the differences between male and female skeletal cases. Interestingly, the univariate follow-up tests suggest that the multivariate difference is largely explained by the differences in the "size" variables (stature and lower limb length) but not by the crural index. Since the crural index is relatively independent of size, these results are intuitively sensible; size differs across sex, but the capacity for locomotion does not.

The B Main Effect. Differences between the marginal means of the Arctic, Great Plains, and Temperate groups are evaluated by testing the hypothesis

$$H_{0(\mathrm{B})}: \frac{1}{2}(\mu_{a_1b_1} + \mu_{a_2b_1}) = \frac{1}{2}(\mu_{a_1b_2} + \mu_{a_2b_2}) = \frac{1}{2}(\mu_{a_1b_3} + \mu_{a_2b_3}) \cdot$$

Table 6.9 Hypotheses SSCP Matrices for A, B, and A × B MANOVA Effects

$$\mathbf{Q}_{H_{(A)}} = \begin{bmatrix} 1900.780 & 110.164 & 72.624 \\ 110.164 & 6.385 & 4.209 \\ 72.624 & 4.209 & 2.775 \end{bmatrix}$$

$$\mathbf{Q}_{H_{(B)}} = \begin{bmatrix} 896.610 & 86.649 & 455.782 \\ 86.649 & 8.633 & 43.761 \\ 455.782 & 43.761 & 232.008 \end{bmatrix}$$

$$\mathbf{Q}_{H_{(A\times B)}} = \begin{bmatrix} 72.810 & 4.743 & 3.576 \\ 4.743 & 0.419 & -0.026 \\ 3.576 & -0.026 & 0.782 \end{bmatrix}$$

$$\mathbf{Q}_{E} = \begin{bmatrix} 2874.967 & 197.43 & -11.529 \\ 197.434 & 74.713 & 0.059 \\ -11.529 & 0.059 & 194.100 \end{bmatrix}$$

The three terms involved in the hypothesis require two row vectors of $\mathbf{L}_B$ to define the contrasts on $\hat{\mathbf{B}}$ and to estimate the hypothesis SSCP matrix. This contrast matrix, with $q_{hB} = 2$, simultaneously compares the means of Arctic versus Great Plains groups and Great Plains versus Temperate groups. These simultaneous contrasts exhaust the $G_B - 1 = 3 - 1 = 2$ available degrees of freedom for the B main effect. The contrast matrix is

$$\mathbf{L}_B = \frac{1}{2}\begin{bmatrix} 1 & -1 & 0 & 1 & -1 & 0 \\ 0 & 1 & -1 & 0 & 1 & -1 \end{bmatrix},$$

which implies an hypothesis of two simultaneous sets of mean differences,

$$H_0\colon \mathbf{L}_B\mathbf{B} = \begin{bmatrix} \frac{1}{2}\left(\mu_{a_1b_1} + \mu_{a_2b_1}\right) - \frac{1}{2}\left(\mu_{a_1b_2} + \mu_{a_2b_2}\right) \\ \frac{1}{2}\left(\mu_{a_1b_2} + \mu_{a_2b_2}\right) - \frac{1}{2}\left(\mu_{a_1b_3} + \mu_{a_2b_3}\right) \end{bmatrix} = \begin{bmatrix} \mathbf{0} \\ \mathbf{0} \end{bmatrix}.$$

Table 6.10 Multivariate Test Statistics on the A, B, and A × B MANOVA Effects

Hypothesis	*Test*	*Test Statistic*	R^2	ν_h, ν_e	*F-approx.*	*p*
A Main Effect	Pillai's *V*	0.406	.406	3, 137	31.16	<.001
	Wilks' Λ	0.594	.406	3, 137	31.16	<.001
	Hotelling's T	0.682	.406	3, 137	31.16	<.001
	Roy's *θ*	0.406	.406	3, 137	31.16	<.001
B Main Effect	Pillai's *V*	0.611	.306	6, 276	20.21	<.001
	Wilks' Λ	0.392	.374	6, 274	27.23	<.001
	Hotelling's T	1.540	.435	6, 272	34.91	<.001
	Roy's *θ*	0.606	.606	3, 138	70.63	<.001
A × B Interaction	Pillai's *V*	0.031	.015	6,276	0.72	.637
	Wilks' Λ	0.969	.016	6, 274	0.71	.639
	Hotelling's T	0.031	.015	6, 272	0.71	.641
	Roy's *θ*	0.026	.026	3, 138	1.22	.304

Note: The values of R_m^2 are partialled for the remaining effects in the model.

Table 6.11 Univariate Follow-Up Test Statistics on A, B, and A × B MANOVA Effects

Hypothesis	*Response Variable*	R^2	df_h, df_e	F	p
A Main Effect	Stature	.398	1, 139	91.90	<.001
	Lower limb length	.079	1, 139	11.88	.001
	Crural index	.014	1, 139	1.99	.161
B Main Effect	Stature	.238	2, 139	21.68	<.001
	Lower limb length	.104	2, 139	8.03	.001
	Crural index	.544	2, 139	83.07	<.001
A × B Interaction	Stature	.025	2, 139	1.76	.176
	Lower limb length	.006	2, 139	0.39	.678
	Crural index	.004	2, 139	0.28	.756

Note: The values of R^2 are partialled for the remaining effects in the model.

For the sample means of $\hat{\mathbf{B}}$, the contrast computes the differences of the unweighted means on the variables of stature, lower limb length, and crural index:

$$H_0 : \mathbf{L}_{\text{B}}\hat{\mathbf{B}} = \begin{bmatrix} -6.04 & -0.54 & -3.12 \\ -2.09 & -0.34 & -0.91 \end{bmatrix}.$$

Testing the hypothesis on the B main effect is accomplished by following the same computational procedures as were followed for the A main effect. With $q_{h\text{B}} = 2$, the hypothesis degrees of freedom for the B main effect are $\nu_h = p * q_{h\text{B}} = 6$. We use $\mathbf{L}_{\text{B}}\hat{\mathbf{B}}$ to obtain $\mathbf{Q}_{H_{(\text{B})}}$ and combined with $\mathbf{Q}_E$ the multivariate test statistics, and the univariate follow-up tests are computed as before. The hypothesis SSCP is presented in Table 6.9, and test statistics are summarized in Tables 6.10 and 6.11.

The multivariate tests on the geographic group differences main effect clearly document the fact that the mean vector of the three response variables differs significantly across groups with about 31% of the joint variation in the stature, lower limb length, and crural index accounted for by differences between the three groups. The univariate follow-up tests on Factor B reveal that while all three of the dependent variables contribute to the multivariate difference, the crural index is most dominant in this regard; its univariate $R^2 = .54$ is substantially larger than for the remaining "size" variables. Judicious follow-up tests of pairwise differences may reveal a further refinement to the meaning of the differences in geographic groups.[10]

The A × B Interaction. The analysis of the interaction between Factors A and B is predicated on testing the hypothesis that the differences between the means of the levels of one factor, say male versus female, will be constant across all levels of the remaining factor, that is across Arctic, Great Plains, and Temperate groups of this example. Most authors recommend that the evaluation and testing of the interaction hypothesis take precedence over the main effects (Muller & Fetterman, 2002, Chap. 14). If the interaction is significant, ignore the main effects (although they must be estimated) and focus attention on the simple main effects of factor mean differences within levels of the remaining factor. Conversely, if the interaction is not significant, the main effects can be reevaluated by removing the

[10]Hypotheses on pairwise differences were illustrated in the four-group, one-way MANOVA of the geographic group classification, and we do not repeat them here as many of the conclusions would be similar.

essential multicollinearity of the interaction and reestimating the model with main effects only or resorting to similar strategies (i.e., Type II sums of squares computations discussed in a later section) to separate the main effects tests from the tests of the interaction. To decide on one of these courses of action, the hypothesis of constant mean differences of one factor across levels of a second factor must be tested. Visual representations of different types of interactions illustrating the correspondence between parallel and nonparallel lines on a graph and the magnitude of the test statistic are given in Myers and Well (2003, Chap. 11). Parallel lines of the graph of the interaction indicate equal mean differences of one factor across the levels of the second factor. When mean differences of one factor are not constant across levels of the second factor, lines of the graph will depart from parallelism. The test of an interaction is a test of parallelism as revealed in the graph. The graph of the means for the stature, lower limb length, and crural index variables cross-classified by sex and geographic cluster ANOVA factors are shown in Figure 6.3.

The hypothesis that male-female differences are constant across geographic clusters can be written as

$$H_{0(\mathrm{A}\times\mathrm{B})}\colon (\mu_{a_1b_1}-\mu_{a_2b_1}) = (\mu_{a_1b_2}-\mu_{a_2b_2}) = \left(\mu_{a_1b_3}-\mu_{a_2b_3}\right).$$

Since $\mathbf{B}=\boldsymbol{\mu}_{\mathrm{AB}}$ for cell mean-coded analyses the contrast matrix, with $q_{h_{AB}} = q_{h_A} q_{h_B} = 2$ rows, is required to define the equality of $H_{0(\mathrm{A}\times\mathrm{B})}$. Thus,

$$\mathbf{L}_{\mathrm{B}} = \begin{bmatrix} 1 & -1 & 0 & -1 & 1 & 0 \\ 0 & 1 & -1 & 0 & -1 & 1 \end{bmatrix}$$

with the null hypothesis also specified as function of the contrast matrix and the cell means,

$$H_0\colon \mathbf{L}_{\mathrm{A}\times\mathrm{B}}\mathbf{B} = \begin{bmatrix} \left(\mu_{a_1b_1}-\mu_{a_1b_2}\right)-\left(\mu_{a_2b_1}-\mu_{a_2b_2}\right) \\ \left(\mu_{a_1b_2}-\mu_{a_1b_3}\right)-\left(\mu_{a_2b_2}-\mu_{a_2b_3}\right) \end{bmatrix} = \begin{bmatrix} \mathbf{0} \\ \mathbf{0} \end{bmatrix}.$$

Substituting the estimates of $\mathbf{B}$ into the contrast defines the quantities in $\mathbf{L}_{\mathrm{A}\times\mathrm{B}}\hat{\mathbf{B}}$ and when applied to the example data gives

$$H_0\colon \mathbf{L}_{\mathrm{A}\times\mathrm{B}}\hat{\mathbf{B}} = \begin{bmatrix} -2.52 & -.057 & -.376 \\ -3.05 & -.346 & -.196 \end{bmatrix}.$$

Figure 6.3 Plots of the Sex by Geographic Cluster Interactions for Three Response Variables

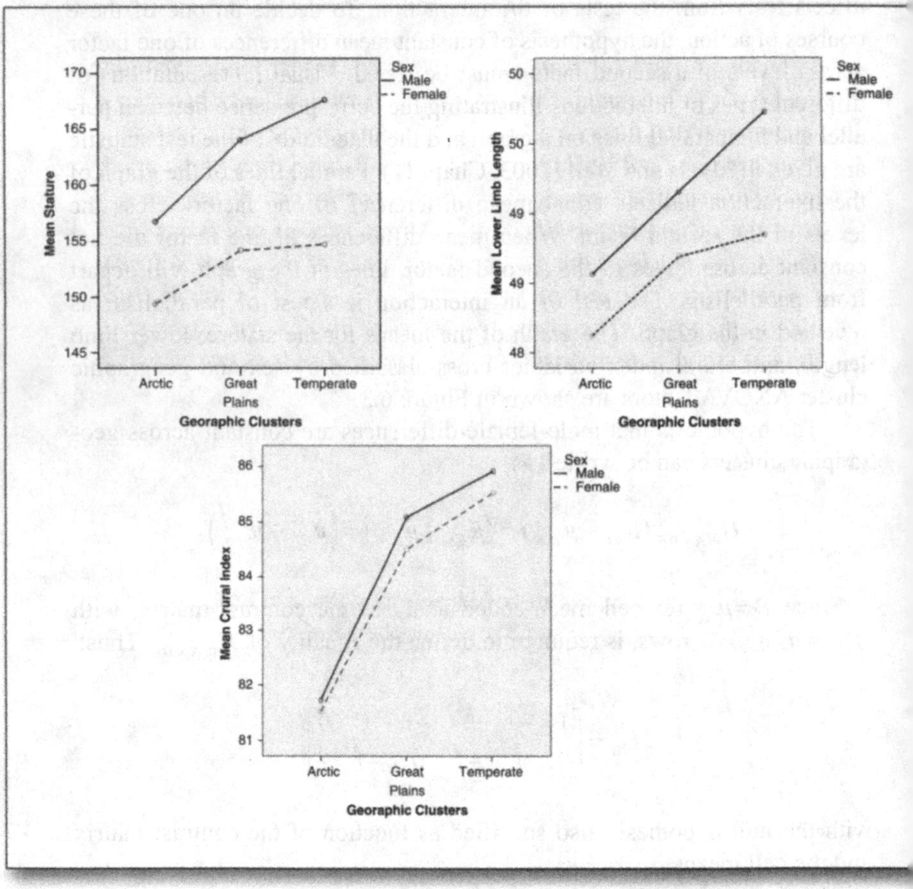

The issue of whether these sample values simultaneously differ significantly from zero can be tested by the same computational routines as were applied to the main effects; we evaluate $\mathbf{Q}_{H_{(A\times B)}}$, use $\mathbf{Q}_E$ from the full model analysis, and evaluate the multivariate and univariate test statistics that depend on $\mathbf{Q}_{H_{(A\times B)}}$. The results of these computations are presented in Tables 6.9, 6.10, and 6.11.

We fail to reject H_0 for the A × B interaction for these data; there is insufficient evidence to conclude that mean differences between males and females are nonconstant across all three geographic clusters. How to proceed in the presence or absence of interaction tests is a complicated matter, and there is no complete agreement as to how to proceed under these two conditions. Many authors (e.g., Muller & Fetterman, 2002, Chap. 14) recommend the strategy of testing and interpreting the interaction effect before attending to the main effects of complex ANOVA models. If the interaction is significant, the main effects should be ignored and interpretation should be placed on the interaction effect and further investigation of the simple main effects that underlie the interaction. Conversely, if the interaction is not significant, it could be removed from the model, and the main effects only model reestimated and interpreted.

A further complication in complex ANOVA designs arises when the data are unbalanced with unequal numbers of observations in the cells of the design as is the case with the 2 × 3 example data used here. Unequal cell sample sizes introduce correlation between the main effects of the design creating nonorthogonal factors in which the effects are no longer independent as would be the case with a balanced design. Several courses of action are possible, including (1) test Factor A, Factor B, and the A × B interaction by adjusting each effect for all other main and interaction effects in the model—this is the Type III sums of squares solution that is a test of unweighted means and which has been used in the previous 2 × 3 example. Type III solutions adjust every effect in a model for all other effects; (2) test the main effects (Factor A or Factor B) adjusted for the other main effect(s) in the model but not adjusted for any higher order term such as the A × B interaction—this is the Type II sums of squares solution that is based on weighted means that adjust for the inequality in cell sample sizes; (3) test the first main effect, say Factor A, unadjusted for any other model effects. Follow by a test of the next main effect, say Factor B, adjusted for Factor A. Next, test the A × B interaction adjusted for both Factors A and B. The idea is to sequentially adjust factors in the model, each effect being adjusted for all preceding effects—this is the Type I sums of squares solution that requires a theory or rationale for choosing the order of the sequence. There is also a Type IV solution that is useful for designs in which one or more cells are empty, but this solution is not recommended by most authors. Detailed discussion of the differences between Type I, II, III, and IV solutions to nonorthogonal designs is given in Green, Marquis, Hershberger, Thompson, and McCollam (1999) and Maxwell and Delaney (2004, Chap. 7). Most statistical software for multivariate linear model analysis defaults to the Type III solution, but the user can request output of other solutions if desired. If the Type III solution is suspect because of markedly unbalanced cell sample size, the most useful alternative would be the Type II solution. In the 2 × 3 MANOVA of the stature estimation data (Table 6.10), a Type II analysis for the A and B main effects requires that the contrast vectors, $\mathbf{L}_A$ and $\mathbf{L}_B$, be constructed to provide weighted averages based on the unequal n_{ab} of the six cells of the design. The method of weighting the contrast vectors by the unequal cell samples sizes to achieve a Type II sums of squares solution is described in detail in Littell, Stroup, and Freund (2002, pp. 198–201). Although we do not present the analysis here, the Type II SSCP analysis of the stature data lead to the same conclusions as those based on the tests summarized in Table 6.10. Most computer software for MANOVA will estimate the parameters of any factorial model by means of a generalized inverse and will print results in the format of the reference cell–coded design matrix discussed in earlier sections. The user may select their preferred analytic method (i.e., Type I-IV partition of the SSCP matrices) for any problem.

A Note on the Assumptions of MANOVA Analyses

The assumptions that are required to justify the test statistics in MANOVA include those assumptions specified for the multivariate linear model in Chapters 3 and 5. Briefly, one can have confidence in the conclusions reached in the analysis if the data meet the following assumptions:

- The relationships in the model $\mathbf{Y} = \mathbf{XB} + \mathbf{E}$ are linear;
- The cases are a random sample drawn from the population about which inferences are to be drawn;
- The observations are independent of one another; there is no systematic influence of the responses of one case on the responses of another;
- The variance-covariance matrix of the errors (or the response variables), as described in Chapters 3 and 5, are constant across cells of the design and are equal to a common population variance-covariance matrix $\mathbf{\Sigma}$—this is the same assumption of constant variance-covariance matrices for the row vectors $\boldsymbol{\epsilon}_i$ (or $\mathbf{y}_i$) and in ANOVA textbooks is described as homogeneity of within-cells variance-covariance matrices. If the $\mathbf{\Sigma}_i$ are constant across all cases, then it follows that they must be constant within cells of the MANOVA design as well. Tests of the equality of variance-covariance matrices across cells of the MANOVA design are given in Rencher (1998, pp. 138–140); and
- The vectors of response variables (and the errors of the model) for each case follows a multivariate normal distribution as described previously. In the MANOVA context, this is assumed to be true within each cell of the design.

More complete details of the assumptions underlying MANOVA and their diagnostic tests are given in Stevens (2007, Chap. 6).

CHAPTER 7. THE EIGENVALUE SOLUTION TO THE MULTIVARIATE LINEAR MODEL: CANONICAL CORRELATION AND MULTIVARIATE TEST STATISTICS

Previous chapters have focused on obtaining multivariate test statistics by the application of traces and determinants of matrix quantities that yield scalar values of the proportion of variance in **Y** associated with **X**. Three of the multivariate test statistics, as presented in Table 4.5, were construed to be direct generalizations of univariate concepts of R^2, $1-R^2$, and the ratio of $\frac{R^2}{1-R^2}$, each based on a function of the hypothesis and error sums of squares and cross-product (SSCP) matrices $\mathbf{Q}_H$ and $\mathbf{Q}_E$. In addition to the test statistics of Pillai, Wilks, and Hotelling, we also introduced, with very brief explanation, a fourth multivariate test statistic, Roy's greatest characteristic root (GCR) test, θ, which we defined as the maximum squared canonical correlation between **Y** and **X**. In this chapter, we introduce the solution to the eigenvalue problem and canonical correlation, which provides yet another perspective on the analysis and interpretation of multivariate linear model problems. In what follows we will see that all four multivariate test statistics derive directly from these eigenvalues. One important advantage of the eigenvalue solution to multivariate analysis is the associated eigenvector that accompanies that solution. The eigenvectors will provide a new tool for understanding the multivariate relationship by the identification of optimal linear combinations of the dependent and predictor variables of **Y** and **X**.

This comprehensive solution to multivariate linear model problems stems directly from the mathematical techniques for solving homogeneous equations and solving for the roots of an *n*th degree polynomial equation. The roots of the polynomial that emerge from the solution to a set of homogeneous equations are the eigenvalues, and the eigenvectors will contain the weights of optimal linear combinations of **Y** and **X** associated with each eigenvalue.

In the remainder of this chapter, we will begin with a conceptual definition of canonical correlation, introduce the eigenvalue problem, and relate the s = minimum[p,q] eigenvalues of $(\mathbf{Q}_E+\mathbf{Q}_H)^{-1}\mathbf{Q}_H$ and $\mathbf{Q}_E^{-1}\mathbf{Q}_H$ to the squared canonical correlations, introduce the optimal linear combinations of **Y** and **X** based on the eigenvectors associated with each eigenvalue, and connect all of the above to the four multivariate test statistics. Finally, we will illustrate these applications on two of the example data sets analyzed in the previous chapters.

A Conceptual Definition of Canonical Correlation

Imagine that we were somehow able to construct a linear combination (l) of a column vector of Y variables and a column vector of weights **a**, say $l = \mathbf{a}'\mathbf{Y} = a_1Y_1 + a_2Y_2 + \ldots + a_pY_p$, and another linear combination (m) of a column vector of X variables and a column vector of weights **b**, say $m = \mathbf{b}'\mathbf{X} = b_1X_1 + b_2X_2 + \ldots b_qX_q$. Since l and m are scalars that can be computed for each case in the data set, we can also compute their variances, S_l^2, S_m^2, and their covariance S_{lm}.[1] The Pearson product-moment correlation coefficient between l and m leads to the conceptual definition of the *canonical correlation*,

$$\rho_{canonical} = \frac{S_{lm}}{\sqrt{S_l^2 S_m^2}}. \qquad [7.1]$$

If we replace l, m, **a**, and **b** with their sample estimates,

$$\hat{l}_1 = \hat{\mathbf{a}}'\mathbf{Y} = \hat{a}_1Y_1 + \hat{a}_2Y_2 + \cdots + \hat{a}_pY_p,$$

$$\hat{m}_1 = \hat{\mathbf{b}}'\mathbf{X} = \hat{b}_1X_1 + \hat{b}_2X_2 + \cdots + \hat{b}_qX_q, \qquad [7.2]$$

then the bivariate correlation between these two *canonical variates* ($\hat{l}_1$ and $\hat{m}_1$) that are constructed by finding two sets of *optimal canonical weights* (the vectors $\hat{\mathbf{a}}' = \hat{a}_1, \hat{a}_2, \ldots, \hat{a}_p$ and $\hat{\mathbf{b}}' = \hat{b}_1, \hat{b}_2, \ldots, \hat{b}_q$) such that the correlation between $\hat{l}_1$ and $\hat{m}_1$ is a maximum is the sample canonical correlation,

$$\hat{\rho}_{\hat{l}_1\hat{m}_1} = \frac{S_{\hat{l}_1\hat{m}_1}}{\sqrt{S_{\hat{l}_1}^2 S_{\hat{m}_1}^2}}. \qquad [7.3]$$

The first canonical correlation, although a maximum, may not completely explain the variance in **Y** as a function of **X**. Successive canonical correlations are possible up to the limit of s = minimum[p,q] as defined in the previous chapters. Each succeeding canonical correlation is constructed to be orthogonal to those that precede it. The second sample canonical correlation is based on two *canonical variates*, $\hat{l}_2$ and $\hat{m}_2$,

---

[1]It can be shown that the variance of the linear combinations $l = \mathbf{a}'\mathbf{Y}$ or $m = \mathbf{b}'\mathbf{X}$ can be obtained by the matrix products $S_l^2 = \mathbf{a}'\mathbf{V}_{YY}\mathbf{a}$ and $S_m^2 = \mathbf{b}'\mathbf{V}_{XX}\mathbf{b}$, and their covariance is given by $S_{lm} = \mathbf{a}'\mathbf{V}_{YX}\mathbf{b}$, where $\mathbf{V}_{YY}, \mathbf{V}_{XX}$, and $\mathbf{V}_{XY}$ are variance-covariance matrices; all three results are scalars for any single linear combination. Throughout the presentation, more extensive technical discussion of the concepts reviewed here can be found in Rencher (1998, Chap. 8, A10, A11).

constructed as optimal linear combinations of a second set of *canonical weights*, $\hat{\mathbf{a}}'$ and $\hat{\mathbf{b}}'$, and the variables **Y** and **X**,

$$\hat{l}_2 = \hat{\mathbf{a}}'\mathbf{Y} = \hat{a}_1 Y_1 + \hat{a}_2 Y_2 + \cdots + \hat{a}_p Y_p,$$

$$\hat{m}_2 = \hat{\mathbf{b}}'\mathbf{X} = \hat{b}_1 X_1 + \hat{b}_2 X_2 + \cdots + \hat{b}_q X_q. \quad [7.4]$$

The sample canonical correlation between $\hat{l}_2$ and $\hat{m}_2$ is

$$\hat{\rho}_{\hat{l}_2 \hat{m}_2} = \frac{S_{\hat{l}_2 \hat{m}_2}}{\sqrt{S^2_{\hat{l}_2} S^2_{\hat{m}_2}}}, \quad [7.5]$$

which is also a maximum. This second pair of canonical variates is estimated subject to the restriction that the correlations between $\hat{l}_2$ and $\hat{l}_1$, $\hat{l}_2$ and $\hat{m}_1$, $\hat{m}_2$ and $\hat{l}_1$, and $\hat{m}_2$ and $\hat{m}_1$ are all zero.

A third, fourth, and further canonical correlations can be similarly computed with all succeeding linear combinations being selected to maximize their correlation while restricting their correlation with all preceding linear combinations to zero. The process continues up to a maximum of $s =$ minimum$[p,q]$ canonical correlations.

Squared Canonical Correlations Are Eigenvalues. The quantity $s =$ minimum $[p,q]$ has been used throughout without giving a good definition of its meaning or its origin. The square of the s canonical correlations $(\hat{\rho}_1^2, \hat{\rho}_2^2, \cdots \hat{\rho}_s^2)$ introduced in preceding paragraphs are the *eigenvalues* of the matrix quantity $\mathbf{R}_{YY}^{-1}\mathbf{R}_{YX}\mathbf{R}_{XX}^{-1}\mathbf{R}_{XY}$ or its equivalent $\left(\mathbf{Q}_E^* + \mathbf{Q}_H^*\right)^{-1} \mathbf{Q}_H^*$. Eigenvalues, also called *latent roots* or *characteristic roots*, are the roots of a polynomial equation of degree s (i.e., quadratic, cubic, quartic, etc). Every eigenvalue has associated with it an *eigenvector*, which contains the canonical weights of the optimal linear combinations $\hat{\mathbf{a}}', \hat{\mathbf{b}}'$ of Equations 7.3 and 7.4. In Chapter 4, it was shown that Pillai's Trace V summarizes the joint variance in **Y** that is associated with the joint variance in **X**. It can be shown that V is also a function of the s squared canonical correlations, which are the eigenvalues of $\left(\mathbf{Q}_E^* + \mathbf{Q}_H^*\right)^{-1} \mathbf{Q}_H^*$ and that two of the remaining multivariate test statistics (Λ and T) can also be defined as a function of the eigenvalues of $\left(\mathbf{Q}_E^* + \mathbf{Q}_H^*\right)^{-1} \mathbf{Q}_H^*$, or of $\mathbf{Q}_E^{*-1}\mathbf{Q}_H^*$.[2] As introduced in the earlier chapters, Roy's GCR, $\theta = \hat{\rho}^2_{max}$, is the maximum squared canonical correlation between **Y**

---

[2]The eigenvalues $\hat{\rho}_i^2$ of $\left(\mathbf{Q}_E^* + \mathbf{Q}_H^*\right)^{-1} \mathbf{Q}_H^*$ and the eigenvalues λ_i of $\mathbf{Q}_E^{*-1}\mathbf{Q}_H^*$ have a straightforward relationship, $\lambda_i = \hat{\rho}_i^2 / (1 - \hat{\rho}_i^2)$, which will be discussed in a later section of this chapter.

and **X**, which is the largest eigenvalue of $\left(\mathbf{Q}_E^* + \mathbf{Q}_H^*\right)^{-1}\mathbf{Q}_H^*$. In what follows, we will provide a connection to the developments of previous chapters by showing how the proportion of shared variance indexed by R_V^2, R_Λ^2, R_T^2, and R_θ^2 are also functions of the s squared canonical correlations.

The Eigenvalues of 2 × 2 Correlation Matrices

The Determinant of 2 × 2 Correlation Matrices. In the context of multivariate linear model analysis, eigenvalues can be thought of as *concentrations of shared variance* within variables that exist in a single correlation or covariance matrix, or between variables that arise in a matrix such as $\left(\mathbf{Q}_E^* + \mathbf{Q}_H^*\right)^{-1}\mathbf{Q}_H^*$, or of $\mathbf{Q}_E^{*-1}\mathbf{Q}_H^*$. Consider for the moment only the upper left quadrant $\mathbf{R}_{YY}$ of a partitioned correlation matrix $\mathbf{R}_{(p+q)\times(p+q)}$ (Equation 4.8) for the variables in **Y** and **X**, and say that $\mathbf{R}_{YY}$ is of order (2 × 2). This matrix might take on three different numeric realizations, labeled $\mathbf{R}_1$, $\mathbf{R}_2$, and $\mathbf{R}_3$ as given by

$$\mathbf{R}_1 = \begin{bmatrix} 1.00 & .01 \\ .01 & 1.00 \end{bmatrix}, \mathbf{R}_2 = \begin{bmatrix} 1.00 & .50 \\ .50 & 1.00 \end{bmatrix}, \mathbf{R}_3 = \begin{bmatrix} 1.00 & .99 \\ .99 & 1.000 \end{bmatrix}.$$

It is clear from each of the three matrices that a different proportion of shared variance is captured in the correlation matrix, that is, $r_{12}^2 = .001$, $r_{12}^2 = .25$, and $r_{12}^2 = .98$, for $\mathbf{R}_1$, $\mathbf{R}_2$, and $\mathbf{R}_3$, respectively. An index of the concentration of shared variance between Y_1 and Y_2 is the determinant of each of the three matrices, $|\mathbf{R}_1| = (1.00)(1.00) - (.01)(.01) = .9999$, $|\mathbf{R}_2| = (1.00)(1.00) - (.50)(.50) = .750$, and $|\mathbf{R}_3| = (1.00)(1.00) - (.99)(.99) = .02$—the smaller the determinant, the greater the concentration of shared variance. Consider the symbolic matrix

$$\mathbf{R}_{(2\times 2)} = \begin{bmatrix} r_{11} & r_{12} \\ r_{21} & r_{22} \end{bmatrix},$$

with determinant of

$$|\mathbf{R}| = \begin{vmatrix} r_{11} & r_{12} \\ r_{21} & r_{22} \end{vmatrix} = r_{11}r_{22} - r_{12}r_{21} = 1 - r_{12}^2, \qquad [7.6]$$

which is the familiar *coefficient of alienation* and is therefore the proportion of variance that is *not* shared by the two variables. Thus, the determinant of a correlation matrix can be used to "squeeze" the many elements of a matrix down into a single scalar that faithfully tracks the relationships embedded in that matrix. As the determinant of these matrices approaches zero, the value of r_{12}^2 approaches 1.00. The determinant is therefore a scalar measure of the *concentration of shared variance*, and it "tracks" the extent to which the variables are related in these matrices since $r_{12}^2 = 1 - \left|\mathbf{R}_{(2\times2)}\right|$. The variance held in common among the variables in $\mathbf{R}_1$, $\mathbf{R}_2$, and $\mathbf{R}_3$ are $1-(1-r_{12}^2)$ = .02, .25, and .98, respectively.[3] Virtually none of the variance of Y_1 and Y_2 is held in common in $\mathbf{R}_1$, while almost all of the variance of Y_1 and Y_2 is overlapping in $\mathbf{R}_3$.

Determinants play a key role in the solution to the eigenvalue problem; extracting the eigenvalues of $\mathbf{R}_{(2\times2)}$, which involves the determinant of certain functions of **R**, will provide a concrete illustration of how those quantities are obtained and eventually used in the solution of multivariate linear model problems.

Extracting the Eigenvalues of **R**. In the abstract, an *eigen-equation* results from a speculation about the relationship between a matrix $\mathbf{R}_{(p\times p)}$, a vector $\mathbf{v}_{(p\times1)}$, and a scalar λ such that

$$\mathbf{Rv} = \lambda\mathbf{v}. \qquad [7.7]$$

If the elements of **R** are known, it is possible to find a vector **v** and a scalar λ that will satisfy Equation 7.7. Equation 7.7 can be restructured by rearranging the terms, $\mathbf{Rv} - \lambda\mathbf{v} = 0$, and including an identity matrix **I** as a conformability multiplier for λ and factoring out **v**. The result for any $p\times p$ matrix is a set of homogeneous equations,[4]

$$\left(\mathbf{R}_{(p\times p)} - \lambda\mathbf{I}_{(p\times p)}\right)\mathbf{v}_{(p\times1)} = \mathbf{0}_{(p\times1)}. \qquad [7.8]$$

Equation 7.8 is often referred to as the *eigen-equation* in which the values of $\lambda_1, \lambda_2, \cdots \lambda_p$ are the eigenvalues of **R**, and the vectors **v** are the eigenvectors associated with each eigenvalue. The eigenvector associated with each eigenvalue consists of a set of weights that define the linear

[3]We should note that since correlation matrices are standardized versions of variance-covariance matrices, the determinant of that correlation matrix can also be called a *generalized variance*, which is the term Wilks (1932) gave to the scalars that capture the concentrations of shared variance of the $\mathbf{Q}_E$ and $\mathbf{Q}_E + \mathbf{Q}_H$ matrices of the multivariate test statistic Λ.

[4]Homogeneous equations are sets of equations whose solution vector is zero as opposed to a set of simultaneous equations whose solution vector is a vector of nonzero constants (i.e., $\mathbf{X'XB} = \mathbf{X'Y}$ and $\mathbf{B} = (\mathbf{X'X})^{-1}\mathbf{X'Y}$). Details are given in Harris (2001, App. C, Derivation D2.12).

combination of the variables that make up the correlation matrix, say Y_1 and Y_2 for this example.

Equation 7.8 has a nontrivial solution (i.e., a solution in which the vector is not arbitrarily equal to zero) only when the matrix $\mathbf{R}_{(p\times p)} - \lambda\mathbf{I}_{(p\times p)}$ is singular,[5] which means that the determinant of this matrix must be set to zero to meet that requirement. Defining the *characteristic equation* as the determinant of $\mathbf{R}_{(p\times p)} - \lambda\mathbf{I}_{(n\times n)}$ set to zero gives

$$\left|\mathbf{R}_{(p\times p)} - \lambda\mathbf{I}_{(p\times p)}\right| = \mathbf{0}. \qquad [7.9]$$

For an **R** matrix of order 2 × 2, this would be written as

$$|\mathrm{R} - \lambda\mathrm{I}| = \left|\begin{matrix} r_{11} & r_{12} \\ r_{21} & r_{22} \end{matrix} - \lambda\begin{pmatrix} 1 & 0 \\ 0 & 1 \end{pmatrix}\right| = 0$$

$$= \left|\begin{matrix} r_{11} & r_{12} \\ r_{21} & r_{22} \end{matrix} - \begin{pmatrix} \lambda & 0 \\ 0 & \lambda \end{pmatrix}\right| = 0$$

$$= \begin{vmatrix} r_{11}-\lambda & r_{12} \\ r_{21} & r_{22}-\lambda \end{vmatrix} = 0 \cdot \qquad [7.10]$$

Expanding the determinant of this 2 × 2 matrix leads to the product,

$$= (r_{11} - \lambda)(r_{22} - \lambda) - (r_{21})(r_{12}) = 0$$

$$= \lambda^2 - r_{11}\lambda - r_{12}\lambda + r_{11}r_{22} - r_{12}r_{21} = 0. \qquad [7.11]$$

For a correlation matrix, $r_{11} = r_{22} = 1$ and $r_{12} = r_{21}$. Making these substitutions and combining the terms leads to the definition of the determinant of Equation 7.9:

$$\lambda^2 - 2\lambda + \left(1 - r_{12}^2\right) = 0. \qquad 7.12$$

Thus, the solution of this characteristic polynomial equation of $\mathbf{R}_{(2\times 2)}$ is a *quadratic equation* in λ with coefficients *a*, *b*, and *c* defining the quadratic $a\lambda^2 - b\lambda + c = 0$. The coefficients of terms in Equation 7.12 are $a = 1$, $b = 2$,

---

[5]A singular matrix has a determinant = 0; if $\mathbf{R} - \lambda\mathbf{I}$ is legitimately equal to zero, then the product $(\mathbf{R} - \lambda\mathbf{I})\,\mathbf{v} = \mathbf{0}$ is a nontrivial solution to the equation set.

and $c = (1 - r_{12}^2)$. The roots, λ_i, of this second-degree polynomial equation are found by evaluating

$$\lambda_i = \frac{-b \pm \sqrt{b^2 - 4ac}}{2a}. \quad [7.13]$$

There are two underlying roots that satisfy the quadratic equation; substituting a, b, and c into Equation 7.13 and solving yields the two values of λ that satisfy the quadratic equation and that are functions of r_{12},

$$\lambda_1 = 1 + r_{12},$$

$$\lambda_2 = 1 - r_{12}. \quad [7.14]$$

The roots of the second-degree polynomial equation, λ_1 and λ_2—the solution to the characteristic Equation 7.9—are the *eigenvalues of* $\mathbf{R}_{(2\times 2)}$ that satisfy Equation 7.9. For a matrix of order 3 × 3, the resulting determinant would be a cubic polynomial, for a 4 × 4 matrix the result would be a quartic polynomial and so on; if the matrix is of the order $p \times p$ the resulting solution would be a polynomial of degree p. When the matrix is a correlation matrix, or one of its relatives such as $(\mathbf{Q}_E^* + \mathbf{Q}_H^*)^{-1}\mathbf{Q}_H^*$, the *eigenvalues are measures of the degree to which the variance of the variables are concentrated in one or more characteristic roots* of the matrix. For the example matrices $\mathbf{R}_1$, $\mathbf{R}_2$, and $\mathbf{R}_3$, the latent roots are summarized in Table 7.1.

A theorem in matrix algebra asserts that the sum of the eigenvalues of a square matrix is equal to the trace of that matrix. Since any $p \times p$ correlation matrix $\mathbf{R}$ consists of 1s on the main diagonal and since a correlation matrix is a variance-covariance matrix of standard scores (e.g., Z_Y, Z_X), the trace of $\mathbf{R}$ defines a summative index of "total variance" in the matrix; that is, $Tr(\mathbf{R}) = p = \Sigma\lambda_i$. It is useful to ascertain the proportion of the variance in the matrix that is concentrated in each of its characteristic roots. Each of these proportions, $\lambda_i/\Sigma\lambda_i$, converted to percentages are the listed in Table 7.1. If the correlation between variables is nearly 1.00 as is the case in $\mathbf{R}_3$, the variance is concentrated almost exclusively in the first latent root. Conversely, if the variables are nearly orthogonal, as in $\mathbf{R}_1$, then the variance of the matrix is distributed equally across all roots. For a modest correlation of $r_{12} = .50$, as in $\mathbf{R}_2$, the majority of the variance is concentrated in the first eigenvalue with the remaining orthogonal variance contained in the second eigenvalue. If there is a single latent root that absorbs a majority of the variance in a matrix, the variance in the matrix is said to be *concentrated*, whereas if the variance in the matrix is more or less equally spread out across all of its latent roots, the pattern is said to be *diffuse*. We will see

Table 7.1 The Eigenvalues of Three Correlation Matrices

	R_1		R_2		R_3	
	λ	*% Trace*	λ	*% Trace*	λ	*% Trace*
$\lambda_1 = 1 + r_{12}$	1.00	50	1.50	75	1.99	99.5
$\lambda_2 = 1 - r_{12}$	1.00	50	.50	25	0.01	0.5
$\Sigma\lambda_i$	2.00	100	2.00	100	2.00	100

momentarily that all four multivariate test statistics can be constructed as a function of certain eigenvalues, and this diffuse-versus-concentrated distinction (Olson, 1974) will become important in distinguishing between test statistics that are constructed from all the eigenvalues (i.e., V, Λ, and T) and those that rely only on the maximum eigenvalue (i.e., θ).

The Eigenvectors of $\mathbf{R}_{(2\times 2)}$

Each eigenvalue of a matrix quantity is associated with *eigenvectors* $\mathbf{v}_i$. An eigenvector is a vector of weights, that when applied to the values of the original variables (Y_1 and Y_2 in this example) of the correlation matrix **R**, are uniquely related to the eigenvalues of that matrix. The coefficients of the eigenvector result from the solution to the homogeneous equations of Equation 7.8; this process can be illustrated using the values of $\mathbf{R}_2$ from Table 7.1 with $r_{12} = .50$ and corresponding eigenvalues of $\lambda_1 = 1 + r_{12} = 1.5$ and $\lambda_2 = 1 - r_{12} = .5$.[6] The values of the vectors $\mathbf{v}_1$ and $\mathbf{v}_2$ associated with the eigenvalues λ_1 and λ_2 are obtained sequentially. Substituting $\lambda_1 = 1.5$ into Equation 7.8 we have

$$\left[\begin{pmatrix} 1.0 & r_{12} \\ r_{21} & 1.0 \end{pmatrix} - 1.5\begin{pmatrix} 1 & 0 \\ 0 & 1 \end{pmatrix}\right]\begin{bmatrix} V_{11} \\ V_{21} \end{bmatrix} = \begin{bmatrix} 0 \\ 0 \end{bmatrix},$$

[6]The symbol λ is used here to be consistent with the previous discussion of the eigenvalues of **R**. It should not be confused with the eigenvalues of $\mathbf{Q}_E^{*-1}\mathbf{Q}_H^*$ leading to Hotelling's T.

which yields two homogeneous equations,

$$\begin{bmatrix} -.5\mathrm{v}_{11} + .5\mathrm{v}_{21} \\ .5\mathrm{v}_{11} - .5\mathrm{v}_{21} \end{bmatrix} = \begin{bmatrix} 0 \\ 0 \end{bmatrix}.$$

Selecting the first row and solving for v_{11} in terms of v_{21} we find

$$\mathrm{v}_{11} = \frac{.5}{.5}\mathrm{v}_{21}$$

and hence $\mathrm{v}_{11} = \mathrm{v}_{21}$. The same solution would have been found by solving the second of the two equations above, since the rows and columns of the matrix are proportional. Because homogeneous equations have an infinite number of proportional solutions, it is customary to "normalize" the vector to unit length by requiring the sums of squares of the values of the vector $\mathbf{v}_i$ to equal 1.00. Ensuring that $\mathbf{v}_i{}'\mathbf{v}_i = 1$ is achieved by dividing each element of $\mathbf{v}_i$ by $1/\sqrt{\Sigma v_i^2}$. The vector $\mathbf{v}_1 = \begin{bmatrix} \mathrm{v}_{11} \\ \mathrm{v}_{21} \end{bmatrix} = \begin{bmatrix} .5 \\ .5 \end{bmatrix}$ would be normalized by premultiplying the vector by the reciprocal of $\sqrt{v_{11}^2 + v_{21}^2}$,

$$\mathbf{v}_1 = \frac{1}{\sqrt{.5^2 + .5^2}} \begin{bmatrix} .5 \\ .5 \end{bmatrix} = \begin{bmatrix} .7071 \\ .7071 \end{bmatrix}.$$

Note that the product $\mathbf{v}_1'\mathbf{v}_1 = 1.00$. The vector contains the unstandardized coefficients of the optimal linear combination (say, f_1) of the variables Y_1 and Y_2 related to $\lambda_1 = 1 + r_{12}$. Thus,

$$f_1 = \mathbf{v}_1(Y_1) + \mathbf{v}_2(Y_2),$$

$$f_1 = .7071(Y_1) + .7071(Y_2).$$

For example, if two $n=5$ vectors of data, say

$$Y_1 = \begin{bmatrix} 1 \\ 2 \\ 3 \\ 4 \\ 5 \end{bmatrix}, \quad Y_2 = \begin{bmatrix} 3 \\ 2 \\ 1 \\ 5 \\ 4 \end{bmatrix},$$

have a correlation of $r_{Y_1Y_2} = .50$, they would give the weighted linear combination of f_1 scores of

$$\begin{bmatrix} 2.828 \\ 2.828 \\ 2.828 \\ 6.364 \\ 6.364 \end{bmatrix}$$

for the five cases. The variance of this vector of scores, $S^2_{f_1} = 3.75$, is equal to the triple product $\mathbf{v}'\mathbf{V}_{YY}\mathbf{v}$ of the eigenvector and the variance-covariance matrix $(\mathbf{V}_{YY})$ of Y_1 and Y_2. If the variables are standardized, then a weighted linear combination in standardized score form could be written as

$$f_{Z1} = .7071(Z_{Y1}) + .7071(Z_{Y2}).$$

The variance of the resulting vector

$$f_{Z_1} = \begin{bmatrix} -0.894 \\ -0.894 \\ -0.894 \\ 1.342 \\ 1.342 \end{bmatrix}$$

is $1.50 = \mathbf{v}'\mathbf{R}_{YY}\mathbf{v} = \lambda_1$—thus the variance of the normalized eigenvector for standard scores is equal to its eigenvalue. It is in this sense that this new derived variable, f_{Z_1}, is directly connected to the eigen-structure of the original correlation matrix $\mathbf{R}_{YY}$ and to the raw data on which it is based.[7]

[7]We should note that if we took one further step in this scaling process and multiplied the vector $\mathbf{V}_1$ by $\sqrt{\lambda_1}$ the resulting vector [.866 .866] would contain the unrotated loadings of the principal components analysis of $\mathbf{R}_2$. The sums of squares of those rescaled loadings would also equal λ_1, and the correlations of Y_1 and Y_2 with either f_l or Z_{f_l} would be .866 and .866. Jollife (2002) gives an extensive account of principal components analysis.

The second of $p = 2$ eigenvectors for this example data is found by substituting the second eigenvalue, $\lambda_2 = 1 - r_{12} = .50$, into Equation 7.8 and solving for its eigenvector $\mathbf{v}_2$. We have

$$\left[\begin{pmatrix} 1.0 & r_{12} \\ r_{21} & 1.0 \end{pmatrix} - .5\begin{pmatrix} 1 & 0 \\ 0 & 1 \end{pmatrix}\right]\begin{bmatrix} v_{21} \\ v_{22} \end{bmatrix} = \begin{bmatrix} 0 \\ 0 \end{bmatrix},$$

performing the arithmetic yields two homogeneous equations with proportional rows or columns for which a solution is found to be

$$\nu_{12} = -\nu_{22}$$

and the eigenvector $\mathbf{v}_2$ contains two values that are equal in magnitude but opposite in sign,

$$\mathbf{v}_2 = \begin{bmatrix} .5 \\ -.5 \end{bmatrix}.$$

Multiplying each element of $\mathbf{v}_2$ by

$$\frac{1}{\sqrt{\Sigma \mathrm{v}^2}} = \frac{1}{.7071}$$

yields the eigenvector for the second characteristic root,

$$\mathbf{v}_2 = \begin{bmatrix} .7071 \\ -.7071 \end{bmatrix}$$

with sum of squares $\mathbf{v}_2{}'\mathbf{v}_2 = 1$. This new derived variable for raw scores is given as

$$f_{Y_2} = .7071(Y_1) - .7071(Y_2)$$

with variance $S^2_{f_2} = 3.75$. For standard scores the function is,

$$f_{Z_2} = .7071\left(Z_{Y_1}\right) - .7071\left(Z_{Y_2}\right)$$

with resulting score vector

$$f_{Z_2} = \begin{bmatrix} .894 \\ .000 \\ .894 \\ -.447 \\ -.447 \end{bmatrix}$$

for the five cases on variables Z_{Y_1} and Z_{Y_2}. The variance of $f_{Z_2} = \mathbf{v}_2'\mathbf{R}_{YY}\mathbf{v}_2 = .50$, which is also the value of λ_2. We have already noted that $\Sigma\lambda_i = p$, which for correlation matrices can be interpreted as the total variance of the variable set. In addition to the fact that the eigenvectors are scaled to unit length, $\mathbf{v}_1{}'\mathbf{v}_1 = \mathbf{v}_2{}'\mathbf{v}_2 = 1$, they are also orthogonal, $\mathbf{v}_1{}'\mathbf{v}_2 = 0$, a feature that guarantees that the linear combinations of the original variables are uncorrelated, $r_{f_1 f_2} = r_{f_{Z_1} f_{Z_2}} = 0$.

The Eigenvalues of $\mathbf{R}_{YY}^{-1}\mathbf{R}_{YX}\mathbf{R}_{XX}^{-1}\mathbf{R}_{XY}$

A sample canonical correlation $\hat{\rho}_{\hat{l}\hat{m}}$ is the correlation between two optimally weighted linear combinations, $\hat{l} = \hat{\mathbf{a}}'\mathbf{Y} = \hat{a}_1 Y_1 + \hat{a}_2 Y_2 + \cdots + \hat{a}_p Y_p$ and $\hat{m} = \hat{\mathbf{b}}'\mathbf{X} = \hat{b}_1 X_1 + \hat{b}_2 X_2 + \cdots + \hat{b}_q X_q$, of two sets of variables $\mathbf{Y}$ and $\mathbf{X}$. The vectors of weights $\hat{\mathbf{a}}$ and $\hat{\mathbf{b}}$ are chosen to maximize the correlation between $\hat{l}$ and $\hat{m}$, where the variance of $\hat{l}$ is $S_{\hat{l}}^2 = \hat{\mathbf{a}}'\mathbf{V}_{YY}\hat{\mathbf{a}}$, and the variance of $\hat{m}$ is $S_{\hat{m}}^2 = \hat{\mathbf{b}}'\mathbf{V}_{XX}\hat{\mathbf{b}}$. In standard score form, $\hat{l} = \hat{\mathbf{a}}'\mathbf{Z}_{\mathbf{Y}}$ and $\hat{m} = \hat{\mathbf{b}}'\mathbf{Z}_{\mathbf{X}}$, then the variances of $\hat{l}$ and $\hat{m}$ are, respectively, $S_l^2 = \hat{\mathbf{a}}'\mathbf{R}_{YY}\hat{\mathbf{a}} = 1$ and $S_m^2 = \hat{\mathbf{b}}'\mathbf{R}_{XX}\hat{\mathbf{b}} = 1$ and the standardized covariance of $\hat{l}$ and $\hat{m}$ is the canonical correlation of Equation 7.3, which can be rewritten as a function of the weight vectors $\hat{\mathbf{a}}$ and $\hat{\mathbf{b}}$ and the correlation matrix $\mathbf{R}_{YX}$,[8]

$$\hat{\rho}_{\hat{l}\hat{m}} = \hat{\mathbf{a}}'\mathbf{R}_{YX}\hat{\mathbf{b}}. \qquad [7.15]$$

[8]Canonical correlation problems can be solved with correlation matrices (**R**.), variance-covariance matrices (**V**), or SSCP matrices (**S**). The eigenvalues and eigenvectors are invariant across all three data structures. We use correlation matrices here to capitalize on the interpretable elements of **R** and the intermediate results discussed in Chapter 4.

The object of this analysis is to choose the vectors $\hat{\mathbf{a}}$ and $\hat{\mathbf{b}}$ that maximize the canonical correlation of Equation 7.15. Finding the maximum of $\hat{\rho}_{lm}$ is a problem in differential calculus in which the partial derivatives of the *squared* canonical correlation of Equation 7.15, subject to the side conditions that $\hat{\mathbf{a}}'\mathbf{R}_{YY}\hat{\mathbf{a}} = \hat{\mathbf{b}}'\mathbf{R}_{XX}\hat{\mathbf{b}} = 1$, are taken with respect to $\hat{\mathbf{a}}$ and then with respect to $\hat{\mathbf{b}}$, and set equal to zero to define the maximum.[9] This procedure results in a system of homogeneous equations whose solution will yield the *eigenvalue* ($\hat{\rho}^2_{lm}$) and the *eigenvector* ($\hat{\mathbf{a}}$) for **Y** by the eigen-equation

$$\left(\mathbf{R}_{YY}^{-1}\mathbf{R}_{YX}\mathbf{R}_{XX}^{-1}\mathbf{R}_{XY} - \hat{\rho}^2\mathbf{I}\right)\hat{\mathbf{a}} = \mathbf{0}. \qquad [7.16]$$

The same eigenvalue ($\hat{\rho}^2_{lm}$), but a different eigenvector (**b**) for **X**, is found by the compliment,

$$\left(\mathbf{R}_{XX}^{-1}\mathbf{R}_{XY}\mathbf{R}_{YY}^{-1}\mathbf{R}_{YX} - \hat{\rho}^2\mathbf{I}\right)\hat{\mathbf{b}} = \mathbf{0}. \qquad [7.17]$$

The eigenvalues that capture the relationships between **Y** and **X** are found by solving either of the two characteristic equations of Equations 7.18,

$$\left|\mathbf{R}_{YY}^{-1}\mathbf{R}_{YX}\mathbf{R}_{XX}^{-1}\mathbf{R}_{XY} - \hat{\rho}^2\mathbf{I}\right| = \mathbf{0} \qquad [7.18]$$

or Equation 7.19,

$$\left|\mathbf{R}_{XX}^{-1}\mathbf{R}_{XY}\mathbf{R}_{YY}^{-1}\mathbf{R}_{YX} - \hat{\rho}^2\mathbf{I}\right| = \mathbf{0}. \qquad [7.19]$$

- Solving for the determinant of the either Equation 7.18 or 7.19 yields a polynomial equation whose order is determined by the smaller of the p response variables in **Y** or the q predictor variables in **X**. The $s = \text{minimum}[p,q]$ nonzero roots of the characteristic polynomial of either Equation 7.18 or 7.19 are the squared canonical correlations between **Y** and **X** and they have several defining features: The eigenvalues are extracted in order of descending magnitude, $\hat{\rho}_1^2 > \hat{\rho}_2^2 > \cdots > \hat{\rho}_s^2$.
- Each successive squared canonical correlation is maximized to explain the residual variance shared between **Y** and **X** after removing the variance of preceding roots.

---

[9]Derivations are given in Harris (2001, Chap. 5, Derivation 5.1).

Eigenvalues can be thought of as measures of *concentrations of shared variance*; in this context, the eigenvalues reflect the extent to which the joint variance of **Y** is shared with **X**, in the sense of that definition contained in Equation 4.19. Some important properties of eigenvalues and eigenvectors include the following:

- The trace of a square matrix, say $\mathbf{R}_{YY}^{-1}\mathbf{R}_{YX}\mathbf{R}_{XX}^{-1}\mathbf{R}_{XY}$, is equal to the sum of its eigenvalues; that is, $Tr\left(\mathbf{R}_{YY}^{-1}\mathbf{R}_{YX}\mathbf{R}_{XX}^{-1}\mathbf{R}_{XY}\right)=\sum_{i=1}^{s}\hat{\rho}_i^2$.
- The determinant of a square matrix, say $\mathbf{R}_{YY}^{-1}\mathbf{R}_{YX}\mathbf{R}_{XX}^{-1}\mathbf{R}_{XY}$, is equal to the product of its eigenvalues; that is, $\left|\mathbf{R}_{YY}^{-1}\mathbf{R}_{YX}\mathbf{R}_{XX}^{-1}\mathbf{R}_{XY}\right|=\prod_{i=1}^{s}\hat{\rho}_i^2$.
- The eigenvectors associated with each eigenvalue, $\hat{l}_1,\hat{m}_1;\hat{l}_2,\hat{m}_2;\cdots;\hat{l}_s,\hat{m}_s$, are orthogonal to all the eigenvectors that precede them.

Eigenvalues, Squared Canonical Correlations, and the Four Multivariate Test Statistics

Pillai's Trace : The Eigenvalues of $\left(\mathbf{Q}_E^*+\mathbf{Q}_H^*\right)^{-1}\mathbf{Q}_H^*$

The quadruple product $\mathbf{R}_{YY}^{-1}\mathbf{R}_{YX}\mathbf{R}_{XX}^{-1}\mathbf{R}_{XY}$ of the characteristic Equation 7.18 on which the solution of the canonical correlation problem depends is the equivalent of $\left(\mathbf{Q}_E^*+\mathbf{Q}_H^*\right)^{-1}\mathbf{Q}_H^*$ since for any standard score full model (where $\mathbf{Q}_F^*=\mathbf{Q}_H^*$) based on q_f predictor variables, $\left(\mathbf{Q}_E^*+\mathbf{Q}_H^*\right)^{-1}=\mathbf{R}_{YY}^{-1}$ and $\mathbf{Q}_H^*=\mathbf{R}_{YX}\mathbf{R}_{XX}^{-1}\mathbf{R}_{XY}$ (Equation 4.19). From these equalities it follows that the eigenvalues of the matrices on which Pillai's Trace V is based are the squared canonical correlations of Equation 7.18. The sample squared canonical correlations of the matrix quantity that defines the between-sets relationship of **Y** and **X** are found by solving the eigen-equation

$$\left[\left(\mathbf{Q}_E^*+\mathbf{Q}_H^*\right)^{-1}\mathbf{Q}_H^*-\hat{\rho}^2\mathbf{I}\right]\hat{\mathbf{a}}=\mathbf{0}. \quad [7.20]$$

The sample squared canonical correlations of Equation 7.18, $\hat{\rho}_1^2>\hat{\rho}_2^2>\cdots>\hat{\rho}_s^2$, are therefore the roots of the characteristic polynomial

$$\left|\left(\mathbf{Q}_E^*+\mathbf{Q}_H^*\right)^{-1}\mathbf{Q}_H^*-\hat{\rho}^2\mathbf{I}\right|=\mathbf{0}. \quad [7.21]$$

Our earlier definition of V (Equation 4.28) involved the trace of a matrix (i.e., $V=Tr[\left(\mathbf{Q}_E+\mathbf{Q}_H\right)^{-1}\mathbf{Q}_H]$). By the theorem in matrix algebra that the sum of the eigenvalues of a matrix is equal to the trace of that matrix, it follows that the multivariate test statistic Pillai's Trace V is also the sum (for $i=1,\cdots,s$) of the squared canonical correlations (the eigenvalues) of Equation 7.18 or 7.21,

$$V = \Sigma\hat{\rho}_i^2. \qquad [7.22]$$

From the definition of Equation 4.30, it follows that the proportion of joint variance in **Y** accounted for by **X** is the arithmetic average of the squared canonical correlations

$$R_V^2 = \frac{\Sigma\hat{\rho}_i^2}{s}. \qquad [7.23]$$

The *F*-test approximation written on the sums of the squared canonical correlations produces a result that is identical to the *F*-test approximation of Equation 5.4

$$F_{(\nu_h,\nu_e)} = \frac{\Sigma\hat{\rho}_i^2}{s - \Sigma\hat{\rho}_i^2} \cdot \frac{\nu_e}{\nu_h} \qquad [7.24]$$

with ν_h and ν_e defined in Equation 5.5.

Wilks' Λ: The Eigenvalues of $\left|\mathbf{I} - \mathbf{R}_{YY}^{-1}\mathbf{R}_{YX}\mathbf{R}_{XX}^{-1}\mathbf{R}_{XY}\right|$

The test statistic Wilks' Λ and its measure of association, R_Λ^2, were introduced in Chapter 4 (Table 4.5 and Equation 4.31) as a ratio of two determinants,

$$\Lambda = \frac{|\mathbf{Q}_E|}{|\mathbf{Q}_E + \mathbf{Q}_H|}.$$

Λ was described as the multivariate generalization of the univariate ratio $\frac{SS_{ERROR}}{SS_{TOTAL}}$, which is related to the squared multiple correlation coefficient by

$$\frac{SS_{ERROR}}{SS_{TOTAL}} = 1 - R^2.$$

As the multivariate analogue of this univariate measure we defined

$$\Lambda = \frac{|\mathbf{Q}_E|}{|\mathbf{Q}_E + \mathbf{Q}_H|} \cong 1 - R_\Lambda^2.$$

Because Λ is a multiplicative product series, its average R_Λ^2, based on the smaller number of variables in p or q, estimates the proportion of variance in **Y** *not* accounted for by **X**. Hence an adjusted multivariate measure of association based on the geometric mean of Λ was given as $R_\Lambda^2 = 1 - \Lambda^{\frac{1}{s}}$. In Chapter 4, we also showed how R_Λ^2 is the geometric mean of a series of fully partialled univariate values of $1 - R^2$.

To understand Λ as a function of the eigenvalue problem, note that the expression $\dfrac{|\mathbf{Q}_E|}{|\mathbf{Q}_E + \mathbf{Q}_H|}$ can be written as $\left|\mathbf{Q}_E^* + \mathbf{Q}_H^*\right|^{-1}\left|\mathbf{Q}_E^*\right|$. Because the product of two determinants is equal to the determinant of the product, an equivalent expression for Λ can be written as

$$\Lambda = \left|\left(\mathbf{Q}_E^* + \mathbf{Q}_H^*\right)^{-1}\mathbf{Q}_E^*\right|. \qquad [7.25]$$

Recalling that $\mathbf{Q}_E^* + \mathbf{Q}_H^*$ in the metric of correlation matrices is $(\mathbf{R}_{YY} - \mathbf{R}_{YX}\mathbf{R}_{XX}^{-1}\mathbf{R}_{XY}) + \mathbf{R}_{YX}\mathbf{R}_{XX}^{-1}\mathbf{R}_{XY} = \mathbf{R}_{YY}$, then on substitution into Equation 7.25, we have the matrix equivalent of $1 - R^2$,

$$\left|\mathbf{R}_{YY}^{-1}(\mathbf{R}_{YY} - \mathbf{R}_{YX}\mathbf{R}_{XX}^{-1}\mathbf{R}_{XY})\right|,$$

$$\left|\mathbf{I} - \mathbf{R}_{YY}^{-1}\mathbf{R}_{YX}\mathbf{R}_{XX}^{-1}\mathbf{R}_{XY}\right|. \qquad [7.26]$$

Thus, the determinant of Equation 7.26 is Wilks' Λ and is seen to be a function of the multivariate equivalent of $1 - R^2$.[10] A theorem in matrix algebra states that the *product* of the eigenvalues of a matrix is equal to the determinant of that matrix. The characteristic equation associated with Λ is given as

$$\left|(\mathbf{I} - \mathbf{R}_{YY}^{-1}\mathbf{R}_{YX}\mathbf{R}_{XX}^{-1}\mathbf{R}_{XY}) - \left(1 - \hat{\rho}_i^2\right)\mathbf{I}\right| = 0, \qquad [7.27]$$

with eigenvalues of 1 minus the squared canonical correlations. By the theorem on determinants, it follows that Wilks' Λ is a product function of the squared canonical correlations of Equation 7.27,

$$\Lambda = \prod_{i=1}^{s}(1 - \hat{\rho}_i^2). \qquad [7.28]$$

[10]Note that if $p = 1$ and $q = 1$, then Equation 7.26 $= 1 - r_{YX}^2$, and if $p = 1$ and $q > 1$, then the determinant in Equation 7.26 $= 1 - R_{Y \cdot X_1 X_2 \cdots X_q}^2$. Since the determinant is a scalar, Equation 7.26 has the same character.

Since Λ is a product of s characteristic roots, the geometric mean of the eigenvalues is $\Lambda^{\frac{1}{s}}$. Subtracting $\Lambda^{\frac{1}{s}}$ from 1 leads to the classical definition of multivariate R^2_Λ as given in Cramer and Nicewander (1979),

$$R^2_\Lambda = 1 - \Lambda^{\frac{1}{s}}. \qquad [7.29]$$

The statistical significance of Λ or R^2_Λ is evaluated by the F-test approximation given in Equations 5.6 to 5.8.

Hotelling's Trace: The Eigenvalues of $Q_E^{-1}Q_H$

Hotelling's Trace T is the third multivariate test statistic that can be computed from the squared canonical correlations. Recall that T is based on a multivariate analogue of the ratio $\frac{SS_{HYPOTHESIS}}{SS_{ERROR}}$. As explained in Chapter 4, this ratio is conceptually equivalent to the matrix formulation $\mathbf{Q}_E^{-1}\mathbf{Q}_H$, which we have seen is approximately equal to the ratio $\frac{R^2_T}{1-R^2_T}$. Hotelling's Trace was defined as

$$\mathrm{T} = Tr[\mathbf{Q}_E^{-1}\mathbf{Q}_H].$$

Substituting the correlation metric equivalents of $\mathbf{Q}_E^{*-1}\mathbf{Q}_H^*$, Hotelling's T is also defined in terms of correlation matrices of the canonical correlation problem,

$$\mathrm{T} = Tr\left[\left(\mathbf{R}_{YY} - \mathbf{R}_{YX}\mathbf{R}_{XX}^{-1}\mathbf{R}_{XY}\right)^{-1}\mathbf{R}_{YX}\mathbf{R}_{XX}^{-1}\mathbf{R}_{XY}\right]. \qquad [7.30]$$

Let λ_i be the characteristic roots of $\mathbf{Q}_E^{*-1}\mathbf{Q}_H^*$ and define Hotelling's T as a function of the roots of the characteristic equation

$$\left|\mathbf{Q}_E^{*-1}\mathbf{Q}_H^* - \lambda_i\mathbf{I}\right| = 0. \qquad [7.31]$$

From the theorem that the sum of the eigenvalues of a matrix is equal to the trace of that matrix, summing the s eigenvalues of $\mathbf{Q}_E^{*-1}\mathbf{Q}_H^*$ as given in Equation 7.31 must be the value of Hotelling's Trace T,

$$\mathrm{T} = \Sigma\lambda_i. \qquad [7.32]$$

There is a connection between the eigenvalues of Equation 7.31 and the squared canonical correlations such that one index can be defined as a function of the other. That is,

$$\hat{\rho}_i^2 = \frac{\lambda_i}{1+\lambda_i} \Leftrightarrow \lambda_i = \frac{\hat{\rho}_i^2}{1-\hat{\rho}_i^2}. \qquad [7.33]$$

Thus Hotelling's T can be defined as a function of the squared canonical correlations,

$$\mathrm{T} = \Sigma \frac{\hat{\rho}_i^2}{1-\hat{\rho}_i^2}. \qquad [7.34]$$

The proportion of joint variance in **Y** given **X** based on Hotelling's test statistic is given as

$$R_{\mathrm{T}}^2 = \frac{\mathrm{T}}{\mathrm{T}+s}. \qquad [7.35]$$

Cramer and Nicewander (1979) show that Equation 7.35 is also equal to one minus the harmonic mean of the s values of $\left(1-\hat{\rho}_i^2\right)$. Their measure, called $\hat{\gamma}_3$, is equivalent to Equation 7.35,

$$\hat{\gamma}_3 = R_{\mathrm{T}}^2 = 1 - \frac{s}{\sum_{i=1}^{s}\left(\frac{1}{1-\hat{\rho}_i^2}\right)}. \qquad [7.36]$$

Arithmetic, geometric, and harmonic means will always produce slightly different averages of the same numbers, which partly explains why the values of R_V^2, R_Λ^2, and R_{T}^2 will often produce different numerical values when applied to the same data.

Roy's GCR.

While definitions of Roy's GCR test, θ, asserted in the previous chapters were unexplained, it should now be clear that θ is the maximum eigenvalue of Equation 7.18 or 7.19 *or* the maximum eigenvalue of Equation 7.31. Some programmers (SPSS, STATA) prefer the maximum squared canonical correlation of Equation 7.18 (ρ_{max}^2) as the definition of θ, whereas others (SAS) prefer the maximum eigenvalue of Equation 7.31 (λ_{max}). Either is acceptable as ρ_{max}^2 and λ_{max} can easily be recovered from the other. The advantage of defining Roy's θ as ρ^2_{max} rather than λ_{max} is that ρ^2_{max} is both a test statistic and a measure of strength of association.

Example 1: The Eigenvalues and Multivariate Test Statistics of the Personality Data

The four multivariate test statistics, their measures of R^2_m, and their F-test approximations are estimable from the eigenvalues of Equation 7.18 or 7.19. Applying Equation 7.18 to the partitioned **R** matrices of the Personality data of Table 2.1 yields the $s = 3$ squared canonical correlations of $\hat{\rho}^2_1 = .2464, \hat{\rho}^2_2 = .1876,$ and $\hat{\rho}^2_3 = .0777$. Based on the maximum squared correlation, about 25% of the variance of the optimal linear combination of the interview preparation and performance variables is accounted for by Neuroticism, Extraversion, and Conscientiousness. Similarly, about 19% and 8%, respectively, is accounted for by the second and third linear combinations of variables. The contribution of each criterion and predictor variable to the first and succeeding latent roots is presented in the succeeding paragraphs. One would proceed to that step only if the relationship between **Y** and **X** is statistically different from zero. To assess that question, the multivariate test statistics, R^2_ms, and F-test approximations are summarized in Table 7.2.

All four test statistics lead to the same conclusion regarding the full model relationship. Roy's GCR test concentrated in the first and largest characteristic root is excessively liberal.[11] Finding a significant whole model relationship establishes that **Y** and **X** are related in one or more meaningful ways. It remains to be seen if each of the $\hat{\rho}^2_i$ are themselves significantly different from zero, and if so it remains to be seen if the canonical weights of the response and explanatory variables are themselves

Table 7.2 Multivariate Test Statistics, R^2_ms, and F-Test Approximations for the Personality Data

Test Statistic	*Value*	*Multivariate* R^2_m	*df*	*F*	*p*
Pillai's V	.5062	.1687	12, 282	4.77	<.001
Wilks' Λ	.5680	.1718	12, 244	4.84	<.001
Hotelling's T	.6357	.1749	12, 272	4.80	<.001
Roy's θ	.2464	.2464	3, 95	10.35	<.001

[11]The more conservative SAS exact test of Roy's θ also gives a p value < .001.

substantial and significantly different from zero. To pursue these questions, the eigenvectors containing the canonical coefficients $\hat{\mathbf{a}}$ and $\hat{\mathbf{b}}$ must first be defined and estimated.

The Eigenvectors of the Squared Canonical Correlations of $\mathbf{R}_{YY}^{-1}\mathbf{R}_{YX}\mathbf{R}_{XX}^{-1}\mathbf{R}_{XY}$

The s nonzero roots of the characteristic polynomial of Equation 7.18 are the squared canonical correlations between **Y** and **X**, $\hat{\rho}_1^2 > \hat{\rho}_2^2 > \cdots > \hat{\rho}_s^2$. As before, eigenvalues can be thought of as *concentrations of shared variance*; in this context, the eigenvalues reflect the extent to which the joint variance of **Y** and **X** are concentrated in a series of s new ways of combining the response and predictor variables to achieve maximum predictability. The contribution of the variables in **Y** and **X** to this shared variance is documented by the eigenvectors associated with each successive eigenvalue. The eigenvectors are found by the same methods introduced earlier with **R** matrices—substituting each eigenvalue back into Equation 7.16 and solving the resulting homogeneous equations for $\hat{\mathbf{a}}_i$ for the response variables, which is then normalized to unit sums of squares and unit variance for standard scores. The eigenvectors $\hat{\mathbf{b}}_i$ for the predictor variables can be recovered from the solution to Equation 7.17 successively for each the $i = 1, 2, \cdots, s$ eigenvalues, or they can be found by the reciprocal relationships of $\hat{\mathbf{a}}_i$ and $\hat{\mathbf{b}}_i$,

$$\hat{\mathbf{a}}_i = \frac{1}{\sqrt{\hat{\rho}_i^2}} \mathbf{R}_{YY}\mathbf{R}_{YX}\hat{\mathbf{b}}_i \qquad [7.37]$$

and

$$\hat{\mathbf{b}}_i = \frac{1}{\sqrt{\hat{\rho}_i^2}} \mathbf{R}_{XX}\mathbf{R}_{XY}\hat{\mathbf{a}}_i. \qquad [7.38]$$

The successive eigenvectors are orthogonal (i.e., uncorrelated), ensuring that the interpretive value of each pair of eigenvectors is independent of the others. Note that the length of the eigenvectors $\hat{\mathbf{a}}_{(p\times 1)}$ and $\hat{\mathbf{b}}_{(q\times 1)}$ will depend on the number of response and predictor variables in the model, while the number of nonzero eigenvalues of Equations 7.18 and 7.19 is the smaller of p and q.

Example 1: The Eigenvectors of the Personality Data

The relative contribution of each predictor and response variable to understanding the relationships between **Y** and **X** is contained in the optimal linear combinations of both sets of variables—the eigenvectors $\hat{\mathbf{a}}_i$ and $\hat{\mathbf{b}}_i$. Substituting the first eigenvalue $\hat{\rho}_1^2 = .2464$ into Equation 7.16, solving the homogeneous equations for the vector $\hat{\mathbf{a}}_1$, and normalizing it to unit variance yields the optimal linear combination for the standardized Y variables,

$$\hat{l}_1 = -.3605Z_{Y_1} + .7548Z_{Y_2} + .0925Z_{Y_3} + .4771Z_{Y_4}$$

and by Equation 7.17 or 7.38, the vector $\hat{m}_1$ is found as

$$\hat{m}_1 = -.1675Z_{X_1} + 1.0024Z_{X_2} - .1070Z_{X_3}.$$

Calculating a score on $\hat{l}_1$ and $\hat{m}_1$ for each of the 99 participants in the experiment and computing, we find $r_{\hat{l}_1\hat{m}_1} = .4964$ ($r^2_{\hat{l}_1\hat{m}_1} = .2464$)—the conceptual definition of canonical correlation.

To obtain the eigenvector pairs for the remaining two eigenvalues we repeat this process twice; the eigenvalues and eigenvectors for the Personality data are summarized in Table 7.3. Both the raw score and standardized canonical coefficients have been included in the table; both sets of coefficients behave in a similar fashion to regression coefficients, and their interpretation is influenced by all the same factors that affect regression coefficients (e.g., multicollinearity). Since many of the variables used in the social sciences have an arbitrary scale, the standardized coefficients are often preferred.

Structure Coefficients. The structure coefficients of Table 7.3 (labeled "***Cor***" in the table) have been proposed by some authors as an additional method for interpreting the relative importance of response and predictor variables to the meaning of the relationships captured in the canonical correlations between **Y** and **X**. Structure coefficients (named after their factor analytic counterparts) are the zero-order correlations between each response or predictor variable and its parent canonical variate, either $\hat{l}_i$ or $\hat{m}_i$. If the Y variables are indexed as $k=1,2,\ldots,p$, and the canonical variates are indexed as $i=1,2,\ldots,s$, then the $p \times s$ matrix of correlations between each of the Y variables and the s canonical variates is

$$\mathbf{R}_{Y_k l_i} = \begin{bmatrix} r_{Y_1 l_1} & r_{Y_1 l_2} & \cdots & r_{Y_1 l_s} \\ r_{Y_2 l_1} & r_{Y_2 l_2} & \cdots & r_{Y_1 l_s} \\ \vdots & \vdots & \ddots & \vdots \\ r_{Y_p l_1} & r_{Y_p l_2} & \cdots & r_{Y_p l_s} \end{bmatrix}. \qquad [7.39]$$

Table 7.3 Eigenvalues and Eigenvectors of the Canonical Analysis of the Personality Data

	$\hat{\rho}_1^2 = .24644$			$\hat{\rho}_2^2 = .18760$			$\hat{\rho}_3^2 = .07213$		
	Raw	*Stan*	*Cor*	*Raw*	*Stan*	*Cor*	*Raw*	*Stan*	*Cor*
Background	–0.087	–0.360	–0.092	0.105	0.431	0.633	–0.239	–0.985	–0.566
Social	0.152	0.755	0.750	–0.018	–0.090	0.300	0.098	0.487	0.014
Follow-up	0.205	0.093	0.480	1.988	0.895	0.772	0.993	0.447	0.052
Offers	1.363	0.477	0.747	–1.234	–0.432	–0.147	–2.564	–0.898	–0.460
Neuroticism	–0.024	–0.167	–0.246	0.025	0.176	–0.015	0.140	0.992	0.969
Extraversion	0.167	1.002	0.984	–0.039	–0.236	0.100	0.042	0.251	0.148
Conscientiousness	–0.018	–0.107	0.257	0.179	1.071	0.958	–0.002	–0.011	–0.127

Note: Background = background preparation; Social = social preparation; Follow-up = follow-up interviews; Offers = offers tendered; Raw = unstandardized coefficients; Stan = standardized coefficients; Cor = structure coefficients; $\hat{\rho}_i^2$ are the squared canonical correlations.

The values of $\mathbf{r}_{Y_k l_i}$ for the response variables are in the columns labeled "*Cor*" of Table 7.3. Repeating the process for the correlations between the $j=1,2,\ldots,q$ predictor variables and the s canonical variates $\hat{m}_i$ gives the $q \times s$ matrix of structure coefficients for the X variables,

$$\mathbf{R}_{X_j m_i} = \begin{bmatrix} r_{X_1 m_1} & r_{X_1 m_2} & \cdots & r_{X_1 m_s} \\ r_{X_2 m_1} & r_{X_2 m_2} & \cdots & r_{X_1 m_s} \\ \vdots & \vdots & \ddots & \vdots \\ r_{X_q m_1} & r_{X_q m_2} & \cdots & r_{X_q m_s} \end{bmatrix}, \qquad [7.40]$$

which appear in the "*Cor*" columns of the predictor variable rows of Table 7.3. The structure coefficients are efficiently estimated from the within-variable correlation matrices and their respective eigenvectors. For the Y variables,

$$\mathbf{r}_{Y_k l_i} = \mathbf{R}_{YY}\,\hat{\mathbf{a}}_i \qquad [7.41]$$

and for the X variables,

$$\mathbf{r}_{X_j m_i} = \mathbf{R}_{XX}\,\hat{\mathbf{b}}_i. \qquad [7.42]$$

The structure coefficients are frequently referred to as the *canonical loadings*, to distinguish them from the canonical coefficients $\hat{\mathbf{a}}_i$ and $\hat{\mathbf{b}}_i$.

Examination of the standardized canonical coefficients of Table 7.3 for the first characteristic root suggests that the first canonical correlation is principally determined by individuals who are high in Extraversion and low in Neuroticism and Conscientiousness. This cluster of traits is related to two of the four outcomes—social preparation for interviewing and the receipt of job offers. Thus, personality traits of people who are warm, gregarious, assertive, and positive are most strongly connected to the social and interactional aspects of the interviewing process. Conversely, the second optimal linear combination of **Y** and **X** is principally defined by individuals who are high in Conscientiousness; these individuals tend to be those who are invited back for follow-up interviews and who engage in background investigation prior to the face-to-face interview. The third independent way in which **Y** and **X** are combined into an optimal predictive relationship is dominated by Neuroticism, which is characterized by anxiety, fear, hostility, and avoidance is most strongly associated with limited background preparation and relatively few job offers. In these example data, both the standardized canonical coefficients and the structure coefficients lead to the same conclusions. These two sets of indices need not necessarily agree—the canonical coefficients, like regression coefficients, are adjusted for the multicollinearity of both the Y and X variables in the model. In contrast, the structure coefficients are essentially a univariate technique—that is they are zero-order correlations between the canonical variates and each response or predictor variable, unadjusted for the other variables in its set (Rencher, 1988). While keeping this distinction in mind both sets of indices can provide useful interpretive information (Thompson, 1984).

Testing Further Hypotheses on the Canonical Correlations and Canonical Coefficients

Although the substantive interpretation of the three canonical correlations of the Personality data appear to be logically defensible, it is also necessary to provide evidence that the relationships are statistically defensible as well. In problems involving multiple roots, there are multiple possible hypotheses on the canonical correlations and on the canonical coefficients. One important hypothesis speculates that the entire set of squared canonical correlations is zero in the population:

$$H_0 : \rho_1^2 = \rho_2^2 = \rho_3^2 = \ldots = \rho_s^2 = 0, \qquad [7.43]$$

which was tested and rejected by all four multivariate test statistics of Table 7.2.

The omnibus multivariate analyses reviewed thus far constitute the first stage of a more complete analysis of the relationship between the **Y** and **X**. Evaluating how many of the characteristic roots are worthy of further examination and interpretation is accomplished by a succession of additional tests to which we now turn.[12] Questions about the importance of the individual squared canonical correlations, of subsets of canonical correlations, and of the individual canonical weights for the eigenvectors of **Y** and **X** will be presented and illustrated with analyses of the Personality data.

Assessing the Dimensionality of the Characteristic Roots. The tests that address the dimensionality of the eigenvalue space have been variously referred to as *tests of dimensionality* (Gittens, 1985, Chap. 3), dimension reduction analysis (Norusis, 1990, Chap. 3), succeeding canonical correlations (Rencher, 2002, Chap. 11), and partitioned U tests (Harris, 2001, Chap. 4). These tests are based on a series of residualized step-down tests designed to

---

[12]Tests on hypotheses other than the whole model, such as those described in Table 5.1, can also be solved by canonical correlation. The eigenvalues produce the same tests of hypotheses as discussed in Chapter 5, but the canonical analysis yields the additional eigenvectors that are specific to any given partialled model. Details are given in Cohen et al. (2003, Chap. 16).

test successively reduced *sets* of eigenvalues and do not perform well as tests of the individual characteristic roots. Harris (2001) argues that using dimension reduction tests beyond their intended purpose (e.g., dimensionality) for testing individual roots leads to excessive Type I error rates and logical inconsistencies. Better-suited tests of the individual roots are discussed in later sections.

The object of the dimension reduction analysis is to decide if a smaller subset of the s canonical correlations is sufficient to describe the relationships between **Y** and **X**. The dimension-reduction testing procedure begins with a first test evaluating the significance of all s latent roots.[13] The 2nd, 3rd, . . ., sth tests that follow are based on sequential reductions in which the variance of the preceding roots is removed and tests of $s-1$ remaining roots, $s-2$ remaining roots, and so on are performed until the last (smallest) root is tested alone. For the $i = 1$ to s characteristic roots, the series of hypotheses to be tested are

$$H_{0_1}: \rho_1^2 = \rho_2^2 = \cdots = \rho_s^2 = 0,$$

$$H_{0_2}: \rho_2^2 = \cdots = \rho_s^2 = 0,$$

$$\vdots$$

$$H_{0s-1}: \rho_{s-1}^2 = \rho_s^2 = 0,$$

$$H_{0s}: \rho_s^2 = 0. \qquad [7.44]$$

The test of H_{0_1} is a test that Roots 1 through s do not differ from zero, H_{0_2} is a test of Roots 2 through s, and so on to the final test of the sth root. The tests of the hypotheses in Equation 7.44 rely on the squared canonical correlations and Wilks' Λ as a test statistic. Let $\Lambda_1, \Lambda_2, \cdots, \Lambda_s$ be the summary statistic for $H_{0_1}, H_{0_2}, \cdots, H_{0_s}$, such that

$$\Lambda_1 = \prod_{i=1}^{s}(1-\rho_i^2),$$

$$\Lambda_2 = \prod_{i=2}^{s}(1-\rho_i^2),$$

$$\vdots$$

$$\Lambda_s = \prod_{i=s}^{s}(1-\rho_i^2). \qquad [7.45]$$

[13]This is the same test as that of the whole model association discussed in the earlier sections.

The successive values of Λ for each hypothesis can be evaluated by using two different approximate test statistics. The first option is a χ^2 test proposed by Bartlett (1939) based on factorizations of Wilks' Λ,

$$\chi^2_{(df)} = -\left\{(n-1) - \frac{1}{2}(p+q+1)\right\}\log_e \Lambda_k. \qquad [7.46]$$

Letting k = the step of the procedure, Equation 7.46 is asymptotically distributed as χ^2 on $(p-k+1)(q-k+1)$ degrees of freedom for the successive tests. Bartlett's test appears in some computer programs that evaluate eigenvalues (e.g., SPSS DISCRIM). Rao's (1951) F-test approximation on the k successively residualized values of Wilks' Λ appears more widely in programmed procedures (e.g., SAS PROC CANCOR, SPSS MANOVA, and STATA CANON) and the hypotheses of Equation 7.44 are evaluated as an approximate F-test (Rencher, 2002, p. 370) on the k successive values of Λ_k,

$$F_{(\nu_h, \nu_e)} = \frac{1-\Lambda_k^{\frac{1}{d}}}{\Lambda_k^{\frac{1}{d}}} \cdot \frac{df_e}{df_h} \qquad [7.47]$$

with degrees of freedom defined by

$$\nu_h = (p-k+1)(q-k+1),$$

$$\nu_e = td - \frac{1}{2}(p-k+1)(q-k+1),$$

$$t = (n-1) - \frac{1}{2}(p+q+1),$$

$$d = \sqrt{\frac{(p-k+1)^2(q-k+1)^2 - 4}{(p-k+1)^2 + (q-k+1)^2 - 5}}. \qquad [7.48]$$

To illustrate, the dimension reduction analysis for the s = 3 canonical correlations of the Personality data are summarized in Table 7.4 for the squared canonical correlations of .2464, .1876, and .0721.

From previous tests, we know that the collection of all three squared canonical correlations differ significantly from zero. The additional tests of Roots 2 and 3 and of 3 alone are also statistically significant and we conclude that retaining and interpreting all three canonical correlations and their eigenvectors is justified. The last test of this series is a legitimate test of the final eigenvalue, but the preceding lines of Table 7.3 do not reflect

Table 7.4 Dimension Reduction Analysis of the Personality–Job Interview Data

Test of	Λ	*df*	*Approximate F*	*p*
Roots 1 to 3	.5680	12, 243.7	4.84	<.0001
Roots 2 to 3	.7538	6, 186	4.71	.0002
Roots 3 to 3	.9279	2, 94	3.65	.0296

tests of the individual roots. Even if we were unable to reject the null hypothesis for Roots 2 and 3 and for Root 3 alone, this does not necessarily imply that the first root alone would be significantly different from zero. Many of the objections to the dimension reduction tests have been raised because of misinterpretations of this sort (Harris, 2001). Tests of the individual roots, however, can be approached by other procedures.

The Lawley Tests of the Individual Canonical Correlations. A test of the hypothesis that any individual characteristic root has been drawn from a population in which $\rho_i^2 = 0$ can be evaluated by an asymptotically normal critical ratio test given by Lawley (1959; Mardia, Kent, & Bibby, 1979, p. 298). The statistical significance of each of the individual roots can help assess their importance to explaining the relationship between predictors and criteria in a way that is not possible in the dimension reduction analysis. The Lawley test statistics[14] are critical ratio tests defined as

$$Z_\rho = \frac{\hat{\rho}_i - E(\rho_i)}{\sqrt{V(\rho_i)}}. \qquad [7.49]$$

Z_ρ is asymptotically normally distributed if n is sufficiently large and provided that the values of ρ_i^2 and $\rho_i^2 - \rho_{i+1}^2$ are not too near to zero.[15]

[14]The Lawley tests are available in the canonical correlation output of SAS PROC CANCOR.

[15]When the canonical correlations are near zero, or when two adjacent canonical correlations are very nearly equal in value, there is no guarantee that the order of the sample roots is the same as the order of the roots in the population. Hence one cannot be certain which sample test statistic applies to which population eigenvalue.

Table 7.5 Lawley Tests of Individual Canonical Correlations for the Personality–Job Interview Data

Root No.	ρ	*Approximate Standard Error*	Z_ρ	p
1	.49643	.07612	6.13	<.0000
2	.43313	.08207	5.28	<.0000
3	.26857	.09373	2.87	.0040

To illustrate, we evaluate Lawley's critical ratio for each of the three characteristic roots of the Personality data. The canonical correlations, their estimated standard errors, test statistics, and p values have been summarized in Table 7.5.

The null hypothesis is rejected for all three canonical correlations for the Personality data. Hence the interpretations applied to the eigenvectors in earlier paragraphs are supported by these tests of the individual canonical correlations.

Roy's GCR Tests of Individual Canonical Correlations. Although Roy's GCR test is most often interpreted to mean that θ is exclusively a test of the maximum characteristic root of $(\mathbf{Q}_E + \mathbf{Q}_H)^{-1}\mathbf{Q}_H$, the GCR test has also been recommended as a conservative test of the individual roots beyond the first (Gittens, 1985, Chap. 3; Harris, 2001, Chap. 5). Critical values of θ have been tabled by Harris (2001, Table A.5). To apply the GCR test to the individual characteristic roots, the values of $\theta_i = \hat{\rho}_i^2$ should be treated in a step-down fashion with the tabled critical value of θ_i being altered at each step of the testing process. The obtained values of θ_i are the squared canonical correlations of $(\mathbf{Q}_E + \mathbf{Q}_H)^{-1}\mathbf{Q}_H$ and are sequentially evaluated against tabled values of $\theta_\alpha(s, m, N)$ where,[16]

$$s = minimum[p, q] - i + 1,$$

$$m = \frac{1}{2}\left(|p - q| - 1\right),$$

$$N = \frac{1}{2}\left(n - q_f - p - 2\right). \quad [7.50]$$

[16]The value of s is altered to accommodate the sequence of tests; N is a parameter of the distribution of θ; n denotes the sample size.

In this usage, the GCR test first evaluates the significance of the maximum squared canonical correlation. By the union-intersection principle of Roy (1957), the test of the hypothesis $\rho_1^2 = 0$ can also be used as a test of the overall relationship between **Y** and **X**—if the maximum squared canonical correlation is not statistically significant, then the null hypothesis could not be rejected for any of the remaining eigenvalues.[17] If the first root is significant, then the focus shifts to the tests of the remaining individual canonical correlations. Tests of succeeding latent roots have the same form but are evaluated against changing critical values of $\theta_{\alpha(s,m,N)}$ denoted by the changing value of *s* as the roots are serially exhausted. The GCR tests on the three squared canonical correlations of the Personality data are summarized in Table 7.6.

The individual GCR tests lead to the same conclusions as did the Lawley test. The evidence of all of the preceding analyses suggests that all three latent roots should be retained. The first latent root accounts for greater than 47.7% of *V*, while the second (37.1% of *V*) and third root (14.3% of *V*) also appear to be substantial and interpretable.

Biplots of the Eigenvalues of $\mathbf{R}_{Y:X}$. Scatterplot matrices of the Personality data shown in the previous chapters give visual information about the

Table 7.6 GCR Tests of Individual Canonical Correlations for the Personality–Job Interview Data

Root No.	$\hat{\rho}^2$	*s*	$\theta_{.05(s,\,0,\,45.5)}$	$\theta_{.01(s,\,0,\,45.5)}$	*p*
1	.2464	3	.150	.188	<.01
2	.1876	2	.111	.147	<.01
3	.0721	1	.064	.096	<.05

Note: The value of *s* for roots 1, 2, and 3 is 3, 2, and 1, respectively.

[17]Compared with the other three multiple root tests, the omnibus hypothesis based on the maximum GCR test has the greatest power for models in which the shared variance is *concentrated* in a single root (Olson, 1974, 1976), and has lesser power for models where the shared variance is diffused throughout more than one latent root.

relationships between the response and criterion variables, but those plots are limited to pairs of variables. Biplots are graphic presentations of the relationships among sets of variables captured in the principal components (eigenvalues) of a covariance or correlation matrix.[18] A biplot presents information about the variances and covariances (correlations) of

Figure 7.1 Biplot of the $p = 4$ and $q = 3$ Variables in the Personality Data

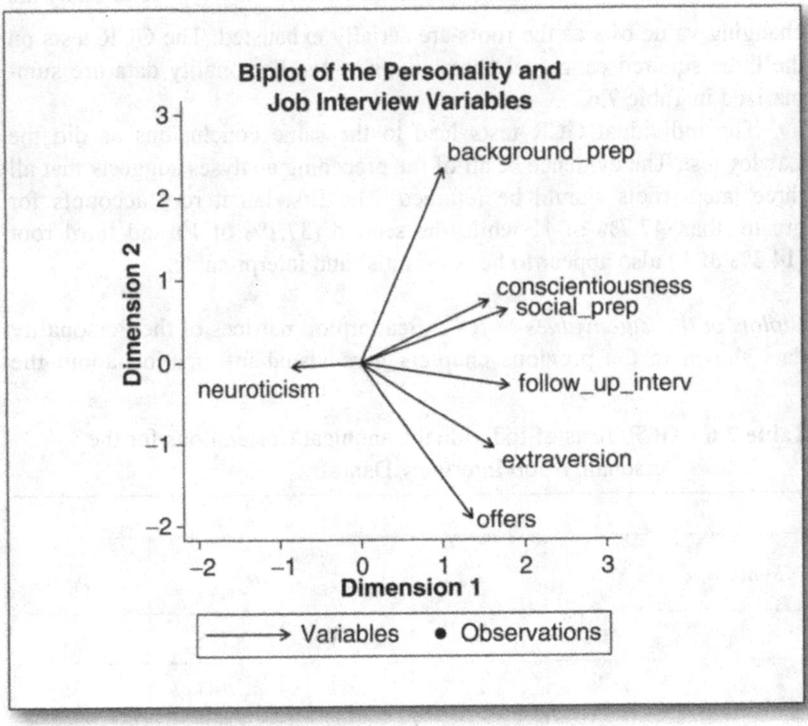

[18]Principal components analysis entails the solution of eigenvalues and eigenvectors of a data matrix (covariance or correlation) and was illustrated in earlier sections of this chapter. The principal components in the biplots shown here are of the covariance matrix containing both **Y** and **X**. The principal components of **Y:X** are a higher order view of the bivariate correlations and therefore an approximation to the principal components of $\mathbf{R}_{YY}^{-1}\mathbf{R}_{YX}\mathbf{R}_{XX}^{-1}\mathbf{R}_{XY}$. The $n = 99$ data points are excluded from this graph since large data sets clutter the display. Biplots are discussed in detail in Gower and Hand (1996).

sets of variables as they are reflected in the principal components (eigenvalues) that underlie a data matrix. For a concatenated data matrix $\mathbf{Y}:\mathbf{X}_{(n\times p+q)}$, the biplot of the first two principal components is shown in Figure 7.1.

The biplot contains information about the variances of the measures (the length of the lines; equal here as the variables are standardized), the correlations between the measures (the cosine of the angle between any two lines), and the data points (deleted from this graph). For the personality data, the biplot suggests considerable common variance between the conscientiousness and both preinterview preparation variables, and similar commonalities between Extraversion and the socially based interview and offer variables. Neuroticism is negatively correlated (angles greater than 180 degrees) with both the predictor and response variables. This visual representation is consistent with the multivariate analyses and univariate follow-up analyses of these data in Chapter 5.

The Regression Tests of the Canonical Coefficients of $l_i = \hat{\mathbf{a}}_i Y$ *and* $m_i = \hat{\mathbf{b}}_i X$. The omnibus multivariate tests of the full set of characteristic roots, the dimension reduction analysis, and the tests on the individual roots can identify the minimally necessary number of underlying canonical variates needed to explain the relationships between **Y** and **X**. One final set of statistical tests that may also aid the interpretive process are tests of the significance of the canonical weights contained in the eigenvectors $\hat{\mathbf{a}}_i$ and $\hat{\mathbf{b}}_i$. Such tests are intended to evaluate the relative importance of each explanatory variable in **X** and each response variable in **Y** to their respective canonical variates l_i and m_i. There is little or no consensus in the literature as to what tests of hypotheses might be suitable for the purpose of evaluating individual canonical coefficients. There are at least two possible alternatives: (1) structural equation model (Bollen, 1989) solutions to canonical correlations, which computes maximum likelihood estimated truly multivariate *t*-tests on the individual coefficients of a MIMIC model (Fan, 1997), and (2) a procedure programmed by Endler and included in STATA CANON (StataCorp., 2007, pp. 65–71) in which canonical variate scores (l_i and m_i) are regressed on their opposite set of variables and the regression coefficients tested by the usual *t* tests.[19] We

[19]The *t*-tests are included as default output of the CANON procedure in STATA Release 10.0 and earlier. As of Release 10.1, the tests are available as an option to the canonical correlation procedure. Because the successive canonical roots are orthogonal, these tests are similar to the Roy-Bargman step-down tests as performed on the successively partialled univariate dependent variables following a significant multivariate test (Stevens, 2009, Chap. 10).

consider only the second alternative here, as it is most proximally related to canonical correlation analysis.

The univariate regression procedure provides a test of a hypothesis on the canonical weights that first requires computing and saving canonical variate scores for each of the *n* cases in the data set and then regressing the canonical variates l_i and m_i on their *opposite* response or predictor variables obtaining univariate squared multiple correlations of $R^2_{l_i \cdot \mathbf{X}}$ and $R^2_{m_i \cdot \mathbf{Y}}$. The values of $R^2_{l_i \cdot \mathbf{Y}}$ and $R^2_{m_i \cdot \mathbf{X}}$ from the regression of the canonical variate scores on their *own* sets of variables (l_i on **Y** and m_i on **X**) are both 1.00 because the canonical variates are perfect linear combinations of their respective variables. However, if we regress each of these optimal linear combinations on the *opposite* set of variables, that is,

$$\hat{l}_i = \hat{\beta}_0 + \hat{\beta}_1 X_1 + \hat{\beta}_2 X_2 + \cdots + \hat{\beta}_q X_q,$$

$$\hat{m}_i = \hat{\gamma}_0 + \hat{\gamma}_1 Y_1 + \hat{\gamma}_2 Y_2 + \cdots + \hat{\gamma}_p Y_p, \qquad [7.51]$$

then the *t*-tests on the regression coefficients $\hat{\beta}$ of l_i regressed on **X** and the coefficients $\hat{\gamma}$ of m_i regressed on **Y** are tests of the significance of the contributions of **Y** to $\hat{m}_i$ and of **X** to $\hat{l}_i$ for the *i*th canonical correlation. From the fitted models of Equation 7.51, the hypotheses $H_0 : \beta_j = 0$, for $j = 1, \cdots, q$ and $H_0 : \gamma_k = 0$, for $k = 1, \cdots, p$ would be tested by the usual *t*-tests on the parameters of these univariate regression models. The values of $R^2_{l_i \cdot \mathbf{X}} = R^2_{m_i \cdot \mathbf{Y}}$ associated with each model of Equation 7.51 are both equal to the squared canonical correlation, $\hat{\rho}^2_{l_i m_i}$, the sum of which for $i = 1, \cdots, s$ canonical variate pairs is Pillai's Trace *V*.[20] These analyses are illustrated here with the Personality data. Since there are three significant roots in the Personality data, three sets of regression models were evaluated and the results are summarized in Table 7.7.

[20]The procedure can be implemented in any software package that allows one to compute the canonical variate scores from the values of the successive eigenvectors of the problem. Any ordinary least squares multiple regression routine can then be used to obtain the test statistics.

Table 7.7 Regression Tests of Variable Contribution to l_i and m_i

	l_1, m_1			l_2, m_2			l_3, m_3		
	Coeff	*t*	*p*	*Coeff*	*t*	*p*	*Coeff*	*t*	*p*
Background	–0.087	–1.70	.089	0.105	1.72	.088	0.239	2.29	.024
Social	0.152	3.53	.001	–0.018	–0.35	.724	–0.098	–1.11	.270
Follow-up	0.205	0.44	.660	1.988	3.59	.001	0.993	–1.04	.301
Offers	1.363	2.27	.025	–1.234	–1.73	.087	2.564	2.09	.040
Neuroticism	–0.024	–0.91	.363	0.025	0.81	.421	–0.140	–2.64	.010
Extraversion	0.167	5.27	<.001	–0.039	–1.04	.300	–0.042	–0.64	.521
Conscientiousness	–0.018	–0.55	.581	0.179	4.66	<.001	0.002	0.03	.978

Note: Background = background preparation; Social = social preparation; Follow-up = follow-up interviews; Offers = offers tendered; Coeff = unstandardized coefficients.

The t-tests on the predictor variables tend to confirm previous interpretations; the first root is largely explained by Extraversion, which predicts social preparation and final offers, the second root is dominated by Conscientiousness, which is most closely related to background preparation and follow-up interviews, and the third root reveals that Neuroticism promotes background preparation but is a serious impediment to receiving job offers. While these t-tests are not strictly speaking a multivariate quantity, they can be informative since the successive eigenvectors $(l_1, l_2, \cdots l_s \text{ and } m_1, m_2, \cdots m_s)$ are orthogonal and therefore independent. If many such tests are performed, it would be advisable to control Type I error rate escalation by a judicious Bonferroni correction. Truly multivariate tests of the statistical significance of the canonical weights are not included in currently available canonical analysis software but can be estimated by structural equation modeling solutions to canonical correlation (Fan, 1997).[21]

Example 2: The PCB-CVD-NPSY Data

The full model relationship between age, gender, and polychlorinated biphenyls (PCB) exposure on memory, cognitive flexibility, and cardiovascular risk factors (cholesterol and triglycerides) investigated in the previous chapters was found to be both substantial ($R_V^2 = .140$) and statistically significant ($p < .0001$). A biplot of those nine variables, shown in Figure 7.2, suggests the possible underlying characteristic roots of the multivariate relationship between **Y** and **X**.

The canonical correlation analysis of these data offers additional insight into these multivariate relationships since the whole model test is identically a function of the $\hat{\rho}_i^2$, that is, $V = \Sigma\hat{\rho}_i^2 = .3538 + .0502 + .0149 = .4189$. The univariate follow-up F-tests (Table 5.4) suggested that both age and PCB exposure play a prominent role in predicting both memory dysfunction and elevated cardiovascular risk factors; neither was seen to affect cognitive flexibility, which was found to be more a function of gender. The dimension reduction analysis reveals that the null hypothesis is rejected for Roots 1 to 3 ($p < .001$) but cannot be rejected for Roots 2 and 3 ($p = .074$), or Root 3 ($p = .429$). In contrast, the Lawley tests and the GCR tests on the individual roots, summarized in Table 7.8, suggest that the first

---

[21]The maximum likelihood estimated t-tests from the SEM MIMIC models fitted to the Personality data and the PCB data produce results that are in good agreement with the t-tests of the univariate regression analyses of l_i and m_i.

Figure 7.2 Biplot of the Nine Variables of the PCB-CVD-NPSY Data

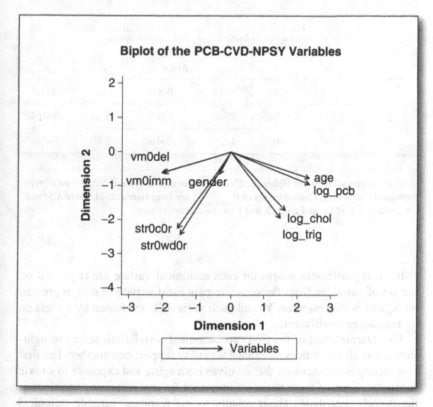

Note: PCB = polychlorinated biphenyls; CVD = cardiovascular disease; NPSY = neuropsychological functioning.

two roots are worthy of further attention, while the third root is probably ignorable even if marginally significant at $\alpha = .05$.

The interpretation of the first two roots is further supported by the fact that they account for 84.5% and 12.0%, respectively, of the trace of V for a total of nearly 97% of the trace; the third root is likely to be of negligible importance despite its statistical significance. The canonical weights for the first two squared canonical correlations are presented in Table 7.9; the eigenvectors are standardized. The column of p values next to each canonical weight are from the univariate procedure where

Table 7.8 Lawley Tests and GCR Tests for the PCB-CVD-NPSY Data

Root No.	*GCR Tests* $\theta_i = \rho_i^2$	p	ρ_i	*Lawley Tests* Standard Error of ρ_i	Z_p	p
1	.3538	<.01	.5948	.0400	14.87	<.0001
2	.0502	<.01	.2240	.0588	3.81	.00014
3	.0149	>.05	.1220	.0610	2.00	.0460

Note: PCB = polychlorinated biphenyls; CVD = cardiovascular disease; NPSY = neuropsychological functioning. Critical values of $\theta_{\alpha(\alpha,s,m,N)}$ are from Harris (2001, Table A.5) and for $m = 2.5$, $N = 125.5$, and $s = 3$, 2, and 1 for the successive roots.

individual participant scores on each canonical variate are regressed on the set of variables from the opposite canonical variate (e.g., l_i regressed on **X**, and m_i regressed on **Y**) and each variable is evaluated by a t-test on its regression coefficient.

The interpretation of the first two canonical correlations seems straightforward as all the indices of Table 7.9 tend to support one another. The first root clearly is a dimension that involves both aging and exposure to a toxic substance. Each of these variables (adjusted for any remaining variables in the model) contributes about equally to the response variable canonical variate, which is dominated by deteriorating memory and increasing risk factors for cardiovascular disease with increases in age and exposure. The second canonical correlation is clearly a gender factor, with influence limited to one of the memory variables—the p values from the follow-up regression analysis test of the response variables reveals the weakness of the relationship. Investigators would be justified in ignoring the third canonical correlation in an interpretation of the data.

In concluding this volume, we reemphasize the fact that canonical correlation analysis subsumes all the analyses presented throughout this volume. A canonical analysis can be performed on any multivariate hypothesis for which the $\mathbf{Q}_H$ of Equation 5.1 can be formulated; consequently canonical analysis can be used to test all the hypotheses in multivariate multiple regression analysis and multivariate analysis of variance, each of which subsumes all of the most common univariate models of the same type (Knapp, 1978). Canonical analysis can also be extended to perform contingency table

Table 7.9 Canonical Weights, Canonical Loadings, and *t*-Tests on the PCB Data

Variable	$\rho_1^2 = .3538$			$\rho_1^2 = .0502$		
	Canonical Weight	*p*	*Canonical Loading*	*Canonical Weight*	*p*	*Canonical Loading*
vmm.imm	–.361	.008	–.670	–.894	.041	–.366
vmm.del	–.262	.055	–.675	.522	.234	–.044
str.w	.023	.856	–.347	.497	.219	.607
str.c	–.260	.039	–.435	.299	.460	.569
cholesterol	.324	.002	.662	.561	.093	.223
triglycerides	.400	<.0001	.653	–.483	.144	–.204
age	.531	<.0001	.921	–.220	.589	.076
gender	–.066	.450	–.111	1.029	<.0001	.974
log_PCB	.539	<.0001	.935	.338	.409	.044

Note: Canonical weights are standardized; vm.imm = immediate visual memory, vm.del = delayed visual memory; str.w = Stroop Word; str.c = Stroop Color.

analyses by analyzing multiple dummy coded variables for both **Y** and **X**. In such analyses, the Pearson χ^2 statistic can be shown to be a function of Pillai's Trace statistic, $\chi^2 = nR_V^2$ (Kshirsagar, 1972, pp. 379–385). We trust that the reader will find many useful applications of canonical correlation and other forms of the multivariate linear model in their own work, and we hope that reading this volume has reinforced the idea that MMR, MANOVA, and CCA are not separate techniques, but rather are an integrated set of techniques all subsumed under the rubric of the general linear model.

REFERENCES

Anderson, T. W. (2003). *An introduction to multivariate statistical analysis*. New York: Wiley.
Arthur, M. M., Van Buren, H. J., & Del Campo, R. J. (2009). The impact of American politics on perceptions of women's golfing abilities. *American Journal of Economics and Sociology, 68,* 517–539.
Auerbach, B. M., & Ruff, C. B. (2010). Stature estimation formulae for indigenous North American populations. *American Journal of Physical Anthropology, 141,* 190–207.
Baek, M. (2009). A comparative analysis of political communication systems and voter turnout. *American Journal of Political Science, 53,* 376–393.
Bartlett, M. (1939). A note on tests of significance in multivariate analysis. *Proceedings of the Cambridge Philosophical Society, 35,* 180–185.
Bollen, K. A. (1989). *Structural equations with latent variables*. New York: Wiley.
Bring, J. (1994). How to standardize regression coefficients. *The American Statistician, 48,* 209–213.
Burdick, R. K. (1982). A note on the multivariate general linear test. *The American Statistician, 36,* 131–132.
Caldwell, D. F., & Burger, J. M. (1998). Personality characteristics of job applicants and success in screening interviews. *Personnel Psychology, 51,* 119–136.
Card, D., Dobkin, C., & Maestas, N. (2009). Does Medicare save lives? *Quarterly Journal of Economics, 124,* 597–636.
Cardoso, H. F. V., & Garcia, S. (2009). The not-so-dark ages: Ecology for human growth in medieval and early twentieth century Portugal as inferred from skeletal growth profiles. *American Journal of Physical Anthropology, 138,* 136–147.
Carpenter, D. O. (2006). Polychlorinated biphenyls (PCBs): Routes of exposure and effects on human health. *Review of Environmental Health, 21,* 1–23.
Cohen, J. (1968). Multiple regression as a general data analytic system. *Psychological Bulletin, 70,* 426–433.
Cohen, J. (1988). *Statistical power analysis for the behavioral sciences*. Mahwah, NJ: Lawrence Erlbaum.
Cohen, J., Cohen, P., West, S. G., & Aiken, L. S. (2003). *Applied multiple regression/correlation analysis for the behavioral sciences*. Mahwah, NJ: Lawrence Erlbaum.
Cohen, J., & Nee, J. C. M. (1984). Estimators for two measures of association for set correlation. *Educational and Psychological Measurement, 44,* 907–917.
Costa, P. T., Jr., & McCrae, R. R. (1992). *Revised NEO personality inventory (NEO-PI-R) and NEO Five Factor Inventory (NEO-FFI) professional manual.* Odessa, FL: Psychological Assessment Resources.
Cramer, E., & Nicewander, A. E. (1979). Some symmetric, invariant measures of multivariate association. *Psychometrika, 44,* 43–54.
Darlington, R. B. (1990). *Regression and linear models*. New York: McGraw-Hill.
Draper, N. R., & Smith, H. (1998). *Applied regression analysis.* New York: Wiley.

Ellis, H. B., MacDonald, H. Z., Lincoln, A. K., & Cabral, H. J. (2008). Mental health of Somali adolescent refugees: The role of trauma, stress, and perceived discrimination. *Journal of Consulting and Clinical Psychology, 76,* 184–193.

Fan, X. (1997). Canonical correlation analysis and structural equation modeling: What do they have in common? *Structural Equation Modeling, 4,* 65–79.

Fox, J. (2009). *A mathematical primer for social scientists*. Thousand Oaks, CA: Sage.

Gittens, R. (1985). *Canonical analysis*. New York: Springer-Verlag.

Glonek, G. F. V., & McCullagh, P. (1995). Multivairate logistic regression. *Journal of the Royal Statistical Society. Series B (Methodological), 57,* 533–546.

Goncharov, A., Haase, R. F., Santiago-Rivera, A., Morse, G. S., Akwesasne Task Force on the Environment, McCaffrey, R. J., et al. (2008). High serum PCBs are associated with elevation of serum lipids and cardiovascular disease in a Native American population. *Environmental Research, 106,* 226–239.

Gower, J. C., & Hand, D. J. (1996). *Biplots*. London: Chapman & Hall.

Green, S. B., Marquis, J. G., Hershberger, S. L., Thompson, M. S., & McCollam, L. M. (1999). The overparameterized analysis of variance model. *Psychological Methods, 4,* 214–233.

Haase, R. F. (1991). Computational formulas for multivariate strength of association from approximate *F* and χ^2 tests. *Multivariate Behavioral Research, 26,* 227–245.

Harris, R. J. (2001). *A primer of multivariate statistics* (3rd ed.). Mahwah, NJ: Lawrence Erlbaum.

Hooper, J. W. (1959). Simultaneous equations and canonical correlation theory. *Econometrica, 27,* 245–256.

Hotelling, H. (1951). A generalized *T*-test and measure of multivariate dispersion. *Proceedings of the Second Berkeley Symposium on Mathematics and Statistics, 23–41.*

Jaccard, J., & Jacoby, J. (2010). *Theory construction and model building skills: A practical guide for social scientists*. New York: Guilford Press.

Jolliffe, I. T. (2002). *Principal components analysis*. New York: Springer.

Kim, J. O., & Ferree, G. (1981). Standardization in causal analysis. *Sociological Methods and Research, 10,* 187–210.

Knapp, T. D. (1978). Canonical correlation as a general data analytic system. *Psychological Bulletin, 85,* 410–416.

Kshirsagar, A. N. (1972). *Multivariate analysis*. New York: Marcel-Dekker.

Lawley, D. N. (1938). A generalization of Fisher's *z*-test. *Biometrika, 30,* 180–187.

Lawley, D. N. (1959). Tests of significance in canonical analysis. *Biometrika, 46,* 59–66.

Lin, K., Guo, N., Tsai, P., Yang, C., & Guo, Y. L. (2008). Neurocognitive changes among elderly exposed to PCBs/PCDFs in Taiwan. *Environmental Health Perspectives, 116,* 184–189.

Littell, R. C., Stroup, W. W., & Freund, R. J. (2002). *SAS system for linear models* (4th ed.). Cary, NC: SAS Institute.

Lubinski, D., & Humphreys, L. G. (1996). Seeing the forest from the trees: When predicting the behavior or status of groups, correlate means. *Psychology, Public Policy and Law, 2,* 363–376.

Mardia, K. V., Kent, J. T., & Bibby, J. M. (1979). *Multivariate analysis*. Amsterdam: Academic Press.

Maxwell, S. E., & Delaney, H. D. (2004). *Designing experiments and analyzing data: A model comparison perspective*. Mahwah, NJ: Lawrence Erlbaum.

Morgan, S. L., & Winship, C. (2007). *Counterfactuals and causal inference. Methods and principles for social research*. Cambridge, UK: Cambridge University Press.

Muller, K. E., & Fetterman, B. A. (2002). *Regression and ANOVA. An integrated approach using SAS software*. Cary, NC: SAS Institute.

Myers, J. L., & Well, A. D. (2003). *Research design and statistical analysis*. Mahwah, NJ: Lawrence Erlbaum.

Namboodiri, K. (1984). *Matrix algebra. An Introduction*. Beverly Hills, CA: Sage.

Norusis, M. J. (1990). *SPSS advanced statistics user's guide*. Chicago: SPSS, Inc.

O'Brien, R. G., & Kaiser, M. K. (1985). MANOVA method for analyzing repeated measures designs: An extensive primer. *Psychological Bulletin, 97,* 316–333.

Olkin, I., & Finn, J. D. (1995). Correlations redux. *Psychological Bulletin, 118,* 155–164.

Olson, C. E. (1974). Comparative robustness of six tests in multivariate analysis of variance. *Journal of the American Statistical Association, 69,* 894–908.

Olson, C. E. (1976). On choosing a test statistic in multivariate analysis of variance. *Psychological Bulletin, 83,* 579–586.

Pekrun, R., Elliot, A. J., & Maier, M. A. (2009). Achievement goals and achievement emotions: Testing a model of their joint relations with academic performance. *Journal of Educational Psychology, 101,* 115–135.

Pillai, K. C. S. (1955). Some new test criteria in multivariate analysis. *Annals of Mathematical Statistics, 26,* 117–121.

Pillai, K. C. S. (1960). *Statistical tables for tests of multivariate hypotheses*. Manila: University of the Philippines Statistical Center.

Puri, M. L., & Sen, P. K. (1971). *Nonparametric methods in multivariate analysis*. New York: Wiley.

Rao, C. R. (1951). An asymptotic expansion of the distribution of Wilks criterion. *Bulletin of the International Statistical Institute, 33,* 177–180.

Rencher, A. C. (1988). On the use of correlations to interpret canonical functions. *Biometrika, 75,* 363–365.

Rencher, A. C. (1998). *Multivariate statistical inference and applications*. New York: Wiley.

Rencher, A. C. (2002). *Methods of multivariate analysis*. New York: Wiley.

Rencher, A. C., & Scott, D. T. (1990). Assessing the contribution of individual variables following rejection of a multivariate hypothesis. *Communications in Statistics: Series B, Simulation and Computation, 19,* 535–553.

Rindskopf, D. (1984). Linear equality restrictions in regression and loglinear models. *Psychological Bulletin, 96,* 597–603.

Robinson, W. S. (1950). Ecological correlations and the behavior of individuals. *American Sociological Review, 15,* 351–357.

Rothman, K. J., Greenland, S., & Lash, T. L. (2008). *Modern epidemiology*. Philadelphia: Wolters Kluwer/Lippincott Williams & Wilkins.

Roy, S. N. (1957). *Some aspects of multivariate analysis*. New York: Wiley.

Schott, J. R. (1997). *Matrix analysis for statistics*. New York: Wiley.

Searle, S. R. (1987). *Linear models for unbalanced data*. New York: Wiley.

StataCorp. (2007). *STATA multivariate statistics reference manual. Release 10*. College Station, TX: StataCorp, LP.

Stevens, J. P. (2007). *Applied multivariate statistics for the social sciences* (2nd ed.). Hillsdale, NJ: Lawrence Erlbaum.

Stewart, D., & Love, W. (1968). A general canonical correlation index. *Psychological Bulletin, 70,* 160–163.

Tatsuoka, M. M. (1988). *Multivariate analysis. Techniques for educational and psychological research*. New York: Wiley.

Thompson, B. (1984). *Canonical correlation analysis: Uses and Interpretations*. Beverly Hills, CA: Sage.

Timm, N. H. (1975). *Multivariate analysis with applications in education and psychology*. Monterey, CA: Brooks/Cole.

Tombaugh, T. N. (2004). Trail Making Test A and B: Normative data stratified by age and education. *Archives of Clinical Neuropsychology, 19,* 203–214.
van den Burg, W., & Lewis, C. (1988). Some properties of two measures of multivariate association. *Psychometrika, 53,* 109–122.
Wilks, S. S. (1932). Certain generalizations in the analysis of variance. *Biometrika, 24,* 471–494.
Zwick, R., & Cramer, E. M. (1986). A multivariate perspective on the analysis of categorical data. *Applied Psychological Measurement, 10,* 141–145.

INDEX

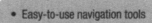

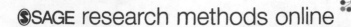

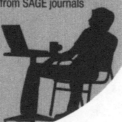

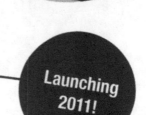

Printed in the United States
By Bookmasters